Signal Processing Applications in CDMA Communications

For a listing of recent titles in the *Artech House Mobile Communications Series,* turn to the back of this book.

Signal Processing Applications in CDMA Communications

Hui Liu

Artech House
Boston • London

Library of Congress Cataloging-in-Publication Data
Liu, Hui.
   Signal processing applications in CDMA communications / Hui Liu.
     p. cm. — (Artech House mobile communications series)
   Includes bibliographical references and index.
   ISBN 1-58053-042-7 (alk. paper)
   1. Code division multiple access. 2. Signal processing. 3. Broadband
communication systems. 4. Mobile communication systems. I. Title. II. Series.

TK5103.452. L58 2000   99-089513
621.3845—dc21   CIP

British Library Cataloguing in Publication Data
Liu, Hui
   Signal processing applications in CDMA communications. —
   (Artech House mobile communications series)
   1. Code division multiple access 2. Wireless communication
   systems 3. Signal processing
   I. Title
   621.3'845

   ISBN 1-58053-042-7

Cover design by Jennifer Stuart

© 2000 ARTECH HOUSE, INC.
685 Canton Street
Norwood, MA 02062

All rights reserved. Printed and bound in the United States of America. No part of this book may be reproduced or utilized in any form or by any means, electronic or mechanical, including photocopying, recording, or by any information storage and retrieval system, without permission in writing from the publisher.
   All terms mentioned in this book that are known to be trademarks or service marks have been appropriately capitalized. Artech House cannot attest to the accuracy of this information. Use of a term in this book should not be regarded as affecting the validity of any trademark or service mark.

International Standard Book Number: 1-58053-042-7
Library of Congress Catalog Card Number: 99-089513

10 9 8 7 6 5 4 3 2 1

Contents

Preface		**ix**
1	**CDMA Fundamentals**	**1**
	1.1 Spread Spectrum	3
	1.1.1 Direct-Sequence Spread Spectrum	3
	1.1.2 Frequency-Hopping Spread Spectrum	6
	1.2 Code-Division Multiple-Access (CDMA)	7
	1.2.1 CDMA Receivers	11
	1.3 Classifications	12
	1.3.1 Short Codes Versus Long Codes	13
	1.3.2 Narrowband Versus Wideband	14
	1.4 Spatial Diversity	19
	1.5 Scope of This Book	21
2	**CDMA With Short Codes: Direct Approaches**	**25**
	2.1 Introduction	25
	2.2 Constrained Optimization	27
	2.3 Decorrelating RAKE Receivers	32
	2.3.1 Constrained MOE	33
	2.3.2 Adaptive Implementation	37
	2.4 Performance Analysis	38
	2.4.1 Receiver Optimality	38
	2.4.2 Steady-State Behavior	41
	2.5 Examples	42
	2.6 Conclusion	46

vi   *Signal Processing Applications in CDMA Communications*

3   CDMA With Short Codes: Indirect Approaches   49
3.1   Introduction 49
3.2   Channel Parameter Estimation 51
   3.2.1   Subspace Concept 52
   3.2.2   Closed-Form Estimation 53
   3.2.3   Overloaded Systems 55
   3.2.4   Performance Analysis 56
3.3   Joint Channel and Carrier Estimation 61
   3.3.1   Problem Reformulation 63
   3.3.2   Decoupling and Closed-Form Solution 66
   3.3.3   Performance Analysis 69
3.4   Examples 72
3.5   Conclusion 73
Appendix 3A: The Existence and Calculation of $\mathbf{G}(z)$ 76
Appendix 3B: First and Second Order Derivatives of the
   $J(\phi, \mathbf{h}, \mathbf{U}_o)$ 78

4   CDMA With Long Codes: Space-Time Processing   81
4.1   Introduction 81
4.2   Uplink 83
   4.2.1   Space-Time Receivers 85
   4.2.2   Blind 2D Reception 90
   4.2.3   Examples 93
4.3   Downlink 101
   4.3.1   Constrained Equalization 103
   4.3.2   Adaptive Implementation 107
   4.3.3   Examples 112
4.4   Conclusion 117

5   Multicarrier CDMA   121
5.1   MC-CDMA Overview 122
   5.1.1   Multicarrier and OFDM 122
   5.1.2   MC-CDMA 126
   5.1.3   MC Direct-Sequence CDMA 130
   5.1.4   Multitone-CDMA 134
5.2   Channel and Carrier Offset Estimation 134

	5.3	Enabling Codes in MC-DS-CDMA 137
	5.4	Carrier Sensitivity Analysis 141
		5.4.1 SINR Degradation 143
	5.5	Discussion 148
	5.6	Conclusion 156
6	**Space-Division Multiple-Access**	**161**
	6.1	Background 162
	6.2	Antenna Array Packet Networks 164
		6.2.1 Conventional Spatial-Division Multiple-Access . 166
	6.3	Scheduling Protocol 170
		6.3.1 Terminal Allocation 170
		6.3.2 Slot Adjustment 173
	6.4	Analysis 176
		6.4.1 Upper-Bound and Lower-Bound 176
		6.4.2 Asymptotic Optimality 178
	6.5	Performance Evaluation 181
		6.5.1 Finite Population With Bursty Traffic 181
		6.5.2 Infinite Population With Poisson Traffic 184
	6.6	Conclusion 186

About the Author   **189**

Index   **191**

Preface

An ideal wireless system provides a unified bandwidth-on-demand platform to support a high user population with mixed traffic and different quality-of-service requirements. Code-division multiple-access (CDMA) is seen as one of the generic signal-access strategies for wireless communications. In addition to many existing applications in mobile telephony, wireless local loops (WLL), and wireless local-area networks (WLAN), CDMA is also positioning itself to become a major player for third-generation personal communication services (PCS) mobile systems.

Signal processing plays a central role in solving many key problems in CDMA, such as diversity combining, multiuser detection, channel estimation, and carrier synchronization. Advances in these areas have received increasing attention in both the wireless communication and signal processing communities. This book is a result of several R&D projects I have been involved in during the past few years. The signal processing methodologies described here target generic wideband CDMA problems in a mobile fading environment. With many books on CDMA focusing on the IS-95 standard, this one should have appeal.

The book consists of six chapters. Chapter 1 provides a preamble that puts the material in subsequent chapters in perspective. Spread spectrum, multiple-access, and different classes of CDMA are included. The chapter highlights some well-known CDMA schemes and discusses their advantages and drawbacks. Readers with a good background on CDMA can skip this part. The chapter also lays the groundwork for subsequent chapters by discussing fading channels, diversity, interference, and other critical issues in CDMA communications. A unified multi-input, multi-output model for antenna array CDMA is established. The model provides a common framework for all discussions in the ensuing chapters.

Chapters 2 to 5 consider wideband CDMA communications over frequency-selective fading channels. In Chapter 2, I discuss CDMA systems with short spreading codes and the advantages of multiuser detection. This chapter focuses on adaptive versions of multiuser detectors and their blind implementation. I describe several techniques that offer significant performance enhancement with complexity comparable to conventional single-user receivers.

Chapter 3 complements Chapter 2 in that it also considers CDMA with short spreading codes. Since operations of some multiuser detectors depend critically on knowledge of spreading waveforms, the principal task here is estimating the fading channel and carrier offsets. I explore the issue of blind signature waveform estimation and present several powerful solutions based on the concept of subspace signal processing.

The primary focus of Chapter 4 is on space-time processing for CDMA with long spreading codes, in particular uplink 2D-RAKE receivers and downlink blind equalizers. Techniques discussed here have direct applications in current CDMA systems such as the IS-95.

Chapter 5 discusses the latest trend in broadband multicarrier CDMA. Most high-speed wireless systems cope with hostile radio channels with highly sophisticated signal processing algorithms. Instead of complicating the system, the technique of multicarrier transmission solves the problem with low-speed, parallel operations. In this chapter, I present a systematic study on multicarrier CDMA and develop a class of high-performance, low-cost modems for broadband multicarrier CDMA communications.

Finally in Chapter 6, I extend the discussion from CDMA to general medium access control (MAC). I introduce the notion of "channel-aware space-division multiple-access," where channel and signal conditions are exploited above the physical layer to achieve throughput multiplication and reduction of packet delays. A new MAC scheduling scheme is introduced to substantially increase the performance of an antenna array packet network.

Much of the material in this book comes from my interactions with my colleagues. Dr. Guanghan Xu at the University of Texas at Austin and Dr. Michael Zoltowski at Purdue University have contributed to the material in this book in many ways. I am also pleased to acknowledge the contributions of my students, Kemin Li, Didem Kivanc, and Hujun Yin. Without their help and ingenuity, this book could not have been written.

Finally, I owe a debt of gratitude to my wife, Qin, my son, Ronan, and my parents for their patience and understanding.

This book was typeset by the author using LaTeX. All simulations and numerical computations were carried out in MATLAB.

Hui Liu
Seattle, Washington

Chapter 1

CDMA Fundamentals

The steadily decreasing cost of wireless communications and its growing deployment have resulted in its increased popularity. In the past decade, wireless personal communication services (PCS) have grown from a vague concept to an important global telecommunication service with over 300 million subscribers. Existing services range from cellular telephony to wireless local loops (WLL) to indoor/outdoor wireless local-area networks (WLAN); see Figure 1.1. Future trends include a wireless infrastructure supporting an integrated mix of multimedia traffic. This unprecedented demand for a new mode of telecommunications has led to a great deal of progress in the past 20 years.

An important challenge in wireless system design is the selection of an appropriate multiple-access scheme. Among many multiple-access strategies that have been developed [1], direct-sequence code-division multiple-access (DS-CDMA) has emerged as a major scheme for current PCS systems and a final candidate for the air interface of the third-generation (3G) universal wireless personal communication network planned for the next century [1, 2]. CDMA radio systems provide several advantages in terms of network planning, graceful degradation under loaded conditions, soft handoff, and path diversity that are fundamental to boosting quality of service and overall capacity.

A vast literature on the topic of CDMA exists, including entire texts on general CDMA [3] and some on the popular IS-95 standard [4]. However, the full potential of CDMA communication, underlined by

2   *Signal Processing Applications in CDMA Communications*

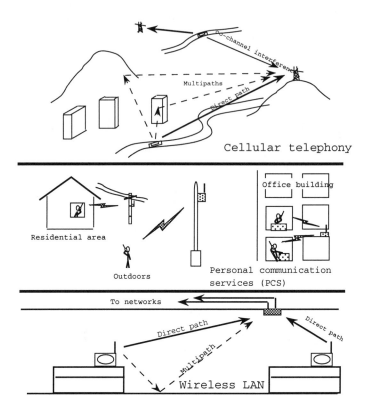

Figure 1.1 Wireless applications.

the continuous growing interest and research activities in this area, is yet to be exploited. The purpose of this chapter is to review the fundamentals of CDMA technology and lay the groundwork for discussion in ensuing chapters. As will be explained in detail in this chapter, CDMA communications rely on spreading the data stream using an assigned spreading code for each user in the time domain. Multiple users transmit their spread signals simultaneously and, in many cases, the received signals are subject to system imperfections, including multipath reflections, carrier offsets, and hardware nonlinearity. How to retrieve the desired signal from self-induced interference and multiple-access interference (MAI) is the key issue in CDMA communications. A number of theoretical and practical questions come, and solutions to these problems offer the promise of significant performance enhancement.

Before proceeding with a thorough description of CDMA, we first introduce the technique of spread spectrum (SS) from which the CDMA technologies are derived. Two types of spread spectrum systems, direct-sequence (DS) SS and frequency-hopping (FH) SS, and their possible applications in commercial wireless communications are described.

The notion of multiple-access will then be established. We present the conventional time-division multiple-access (TDMA) and frequency-division multiple-access (FDMA) schemes and then explore the possibility of code-division multiple-access using DS-SS techniques. After that, we will establish a unified framework under which CDMA is viewed as a generic multiple-access scheme, whereas TDMA and FDMA are merely considered to be special cases of CDMA.

Following the overview of CDMA systems, we introduce a wireless multipath channel model that will be used throughout the book. In developing the channel model, we shall formulate the main problem in CDMA as well as other forms of wireless communications, that is, signal demodulation in the presence of frequency-selective fading channels. Finally, the book ends with a list of issues to be addressed.

1.1 Spread Spectrum

Developed initially for military antijamming communications in the mid-1950s, spread spectrum (SS) has since found a wide range of applications in commercial wireless systems [3]. The underlying idea of spread spectrum is to spread a signal over a large frequency band and transmit it with low power per unit bandwidth. Among many possible ways of spreading the bandwidth, the two predominant types are direct-sequence (DS) and frequency hopping (FH) spread-spectrum.

1.1.1 Direct-Sequence Spread Spectrum

DS spread spectrum achieves band spreading by modulating the information symbol stream with a higher rate chip sequence. Figure 1.2

4   Signal Processing Applications in CDMA Communications

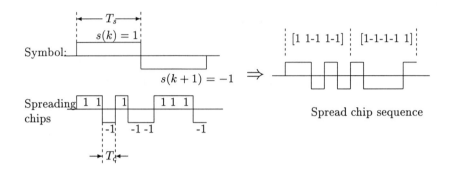

Figure 1.2 Spreading in SS communications.

illustrates the basic principle of a DS spread spectrum digital communications system with a binary information sequence. As seen, each symbol of duration T_s is spread into multiple chips of duration $T_c < T_s$. The bandwidth expansion factor L,

$$L = \frac{T_s}{T_c}$$

determines the amount of redundancy injected during modulation. L is often called the "spreading factor" or the processing gain.

In practice, pseudorandom noise (PN) chip sequences are often employed to make the spread signal as random as possible. The PN sequences can be generated by combining the outputs of feedback shift registers. A well-known class of PN sequences is the maximum length sequences or m-sequence. See [3] for more discussion on PN sequence generation.

After spreading, the chip sequence is usually shaped by a chip pulse shaping filter, $p(t)$, to limit the bandwidth of the output. Mathematically, the spread spectrum modulated signal can be expressed as:

$$x(t) = \sum_{k=-\infty}^{\infty} s(k) w_k(t - kT) \qquad (1.1)$$

where:

$$w_k(t) = \sum_{l=1}^{L} c_k(l) p(t - lT_c + T_c)$$

is the "spreading waveform" or "signature waveform" for the kth symbol. In effect, spread spectrum can be viewed as band excess waveform modulation with the resulting baseband bandwidth on the order of $B_c = 1/T_c$.

The increase in bandwidth or signal dimensionality provides the needed interference/noise resistance in military communications. By the Landau-Pollak theorem, the space of a waveform band-limited to B Hz and approximately time-limited to T sec has approximate dimension of BT [5]. In spread spectrum $B_c T_s = L \gg 1$. The fact that each distinct spreading waveform $w_k(t)$ lies in 1 out of L dimensions makes the spread spectrum signal immune to randomly positioned interfering signals.

At the receiver side, a matched filter captures the desired signal from the one-dimensional subspace defined by the spreading waveform $w_k(t)$ and only those responses to the interference/jammer that lie in the direction of the signal. Effectively, the interference/jammer power is reduced by L. In other words, the spreading gain L measures the degree of freedom of the system and quantifies the interference/noise resistance of a spread spectrum signal.

In practice, it suffices to quantify the signal-to-interference-and-noise ratio (SINR) at the output of the matched filter as a function proportional to L, the processing gain. Theoretically, one can reach arbitrary jamming/interference resistance by expanding the bandwidth.

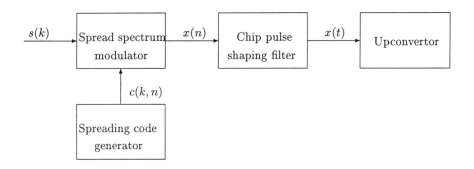

Figure 1.3 A direct-sequence spread spectrum modulator.

The essential modules in a DS-SS modulator are illustrated in Figure 1.3.

1.1.2 Frequency-Hopping Spread Spectrum

Instead of using the code sequences to spread the signal in the time domain, frequency-hopping spread spectrum transmits information through L possible frequency bands based on an index provided by the codes. In particular, an FH spread spectrum signal can be modeled as:

$$s(t) = \sum_{l=1}^{L} c_k(l) e^{j2\pi(l-1)\Delta f t} s(k) p(t - kT)$$

where the binary code, $\{c_k(l) \in [0,1]\}$, determines at which frequency the information is transmitted. The frequency pattern is designed to vary with time, as illustrated in Figure 1.4, where the shaded areas denote the frequency bins through which the information is transmitted at a given time for a given symbol. If only a single-carrier frequency is used on each hop, the data modulation is called "single channel modulation." Frequency-hopping may also be classified as fast or slow. "Fast frequency-hopping" occurs when the hopping rate equals or exceeds the symbol rate, whereas "slow frequency-hopping" implies that many symbols are transmitted in the time interval between frequency hops. The hopping pattern shown in Figure 1.4 is that of a typical fast frequency-hopping with multichannel modulation.

One can regard the FH-SS as the frequency domain counterpart of DS-SS by considering L as the dimensionality of the system – FH-SS transmits its signals using one out of the 2^L possible frequency patterns in an L-dimensional space. At the receiver, an identical code generator controls the frequency synthesizer to remove the pseudo-random frequency pattern before performing regular demodulation. I will elaborate on this concept in the context of multicarrier CDMA in Chapter 6.

In classic military applications, frequency-hopping is particularly efficient in combating the so-called follower jammer, which detects the presence of a signal and sends a jamming signal in the same band. Fast FH prevents the jammer from having sufficient time to intercept the frequency. In commercial applications, FH offers one means to cope with frequency selective fading channels by avoiding the problem of stationary or slowly moving mobile stations being subject to prolonged deep fades. When used properly, FH can also alleviate co-channel interference between neighboring cells by randomizing the co-channel interference pattern. The technique increases the frequency reuse factor in a multiple-cell environment and hence increases the channel capacity of a wireless system.

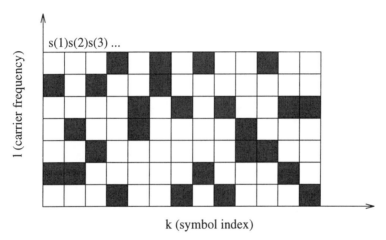

Figure 1.4 The time-varying spectral pattern of an FH spread spectrum signal.

1.2 Code-Division Multiple-Access (CDMA)

In most wireless systems, a central base station simultaneously serves a number of subscribers that share a common transmission medium. This allocation of resources between multiple users is commonly known as multiple access.

All multiple-access techniques require that the messages corresponding to different users be separated in some fashion so that they do not interfere with one another. Conventional techniques accomplish this by making the messages orthogonal in a domain that can be easily handled with digital signal processing (DPS) technologies. In par-

8   *Signal Processing Applications in CDMA Communications*

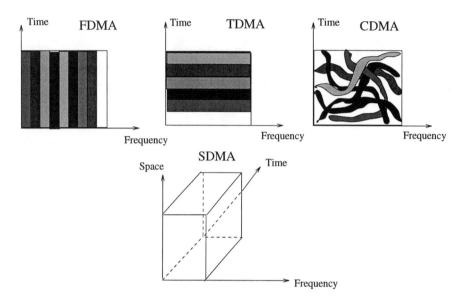

Figure 1.5 Multiple-access schemes.

ticular, time-division multiple-access (TDMA) refers to transmitting different signals at different time slots, whereas in frequency-division multiple-access (FDMA), the base station communicates with different users through different frequency channels. For systems with antenna arrays, the "spatial diversity" can be exploited for multiple-access and the corresponding MA scheme is generally referred to as space-division multiple-access (SDMA). I will talk about SDMA in Chapter 6.

As shown in Figure 1.5, TDMA and FDMA essentially divide a fixed amount of resources – time and frequency – in different ways. Theoretically, an infinite number of possible partitions exists, although most of these are impractical due to various types of implementation difficulties.

The fact that a DS-SS signal can coexist with interference in a high-dimensional space suggests a new way of multiple-access communications. Indeed, instead of assigning different users to different time slots or frequency bins, one can modulate different multiple-access signals with different spreading waveforms occupying one dimension of the total time-frequency resource. With a proper design of the spreading waveforms, we are able to achieve near-orthogonality of waveforms despite the fact that many subscribers share the same spectrum.

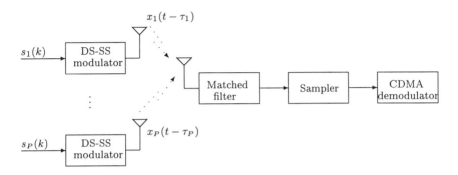

Figure 1.6 A code-division multiple-access scenario.

At the receiver site, multiple-access signals are distinguished by their signature waveforms. Although in a time-frequency plane these spreading waveforms may seem to overlap, they are actually orthogonal or at least linearly independent in the higher dimensional space.

Figure 1.6 depicts a typical CDMA "uplink" scenario (i.e., from the remote subscribers to the base station) where each accessing user modulates its signal using a DS-SS modulator seeded with a pre-assigned spreading code sequence. The received signal is the superposition of all transmitted signals. Thus if the CDMA system has P users, according to (1.1) the received signal can be written as:

$$y(t) = \sum_{i=1}^{P} x_i(t - \tau_i) = \sum_{i=1}^{P} \sum_{k=-\infty}^{\infty} s_i(k) w_{i,k}(t - kT - \tau_i) \qquad (1.2)$$

Here, the subscript i denotes the user index and $\{\tau_i\}$ represents the relative delays of the users' signals.

While the actual implementation may differ, a similar strategy applies to "downlink" (i.e., from the base station to the remote users) where the base station mixes and broadcasts spread spectrum modulated signals designated for different users. Naturally in this case, all signals are synchronized and the relative delays $\{\tau_i\}$ are zero.

10 *Signal Processing Applications in CDMA Communications*

It is worthwhile to point out that despite the apparent differences, CDMA communications can be viewed as a generic multiple-access scheme that encompasses both TDMA and FDMA. This is explained in Figure 1.7, from which we see that:

- If we set the ith user's spreading waveform to be the ones in Figure 1.7 (a), signals from multiple users are non-overlapping in time. The particular CDMA scheme reduces to TDMA, where orthogonality in the time domain is utilized by separate users.

- Similarly, one can regard an FDMA system as a special case of CDMA where eigenfunctions (complex exponentials) are used as the spreading waveforms; see Figure 1.7 (b).

- For the benefits of easy modulation and demodulation, a practical CDMA system often utilizes binary chip waveforms as shown in Figure 1.7 (c).

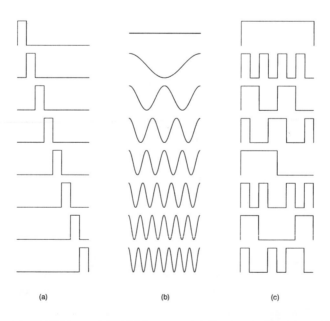

Figure 1.7 TDMA and FDMA as special cases of generic CDMA.

1.2.1 CDMA Receivers

In [6], Verdu derived the optimal CDMA sequence detector that maximizes the conditional probability of a given output sequence. The optimum receiver is comprised of (1) a set of matched filters matched to all P active users' signature waveforms, and (2) a joint detector that maximizes the posterior probability over all possible symbol values. The results were of great theoretical but little practical interest, because the optimum receiver has exponential complexity in the number of users. A large number of suboptimum solutions with lower complexity has been described in the literature since then; most can be classified into two fundamentally different categories, namely, multiuser detectors and single-user detectors.

Multiuser Detectors

A multiuser CDMA receiver has the same matched filter front-end as the optimum receiver. Instead of performing joint maximum likelihood detection, a suboptimum multiuser receiver processes the P dimensional matched filter output vector to arrive at estimates of the corresponding P symbols for respective time intervals. One of the popular multiuser detectors is the linear decorrelating receiver that weighs and combines the matched filter outputs to decouple the superimposed signals based on the signal space defined by their signature waveforms [7–10]. Signal detection can then be implemented individually on the outputs of the decorrelator as illustrated in Figure 1.8.

Among other variations, the minimum mean-squared error (MMSE) receiver selects the combining coefficients, $\{g_{il}\}$, to minimize the metric $E\{|\hat{s}_i(k) - s_i(k)|^2\}$, where:

$$\hat{s}_i(k) = \sum_{l=1}^{P} g_{il}^* z(l)$$

Another popular linear receiver is the zero-forcing receiver that eliminates multiple-access interference.

In practice, when only discrete data samples are available, multiuser detection is performed directly on sample sequences without an explicit form of matched filter. In this book, I loosely define multiuser detectors as those that exploit the algebraic structures of all users' signature waveforms.

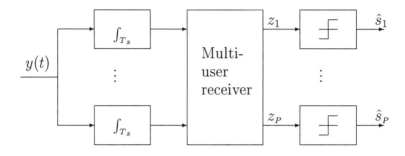

Figure 1.8 Multiuser receiver structure.

Single-User Detectors

In many applications, the detection of the ith user's symbols is based solely on the ith signature waveform; other CDMA signals are treated as unknown interference. This type of receiver is referred to as a single-user detector. Most single-user detectors are of much smaller complexity than multiuser detectors and rely on fewer assumptions. Typically, they assume that the spreading code and timing of the signal-of-interest (SOI) is known. A widely used single-user receiver is the RAKE receiver, which enhances the reception performance by capturing signal energy of the resolvable multipath [11, 12]. Most single-user receivers suffer from the well-known near-far problem when the signal-of-interest is overwhelmed by interfering signals.

1.3 Classifications

Depending on the selection of the spreading waveform $\{w_{i,k}(t)\}$ and the delay spread of multipath channels, a CDMA system can have short- or long-spreading codes and be narrowband or wideband.

1.3.1   Short Codes Versus Long Codes

The spreading codes used in original spread spectrum are pseudo-random in nature. Without a reference, these codes are noise-like, which serves the purpose of antijamming or low probability of intercept communications. In some applications, the period of the spreading codes used in a CDMA system is short relative to the symbol period. When the code period is the same as the symbol duration, the spreading waveform becomes "periodic." That is, the spreading waveform remains the same from symbol to symbol,

$$w_{i,k}(t) = w_i(t), \quad i = 1, \cdots, P$$

The reasons for adopting short spreading codes are, broadly speaking, to enhance the performance of CDMA communication by enabling multiuser detection that requires a time-invariant algebraic structure between users' signature waveforms [7–9, 13–15]. However, due to the complexity of multiuser detection and historical designs of some existing systems, CDMA with short codes seems to be slow to penetrate the commercial world, despite the vast activities in academia that show many fundamental advantages of periodic CDMA. This perspective is changing as computing power becomes cheaper and more plentiful, making some of the algorithms previously deemed too expensive to implement. "Periodic CDMA" or CDMA with short codes is being considered for the third-generation high-performance PCS.

The spreading waveforms for CDMA systems that inherit long spreading codes are "aperiodic," that is, they vary from symbol to symbol. The use of long PN codes randomizes the spectrum of CDMA signals and also alleviates cochannel interference in a multicell setup. Several commercial CDMA systems, including the popular IS-95 standard, employ aperiodic codes. On the downside, the aperiodicity of the spreading waveforms presents a major technical problem in high-performance joint detection since time-varying (from symbol to symbol) multiuser detectors are difficult, if not impossible, to implement.

1.3.2 Narrowband Versus Wideband

In wireless communications, the channel response of a wireless radio with multipath can be modeled as [11]:

$$f(t) = \sum_{l=1}^{L_m} \alpha_l \delta(t - \tau_l) \tag{1.3}$$

where L_m is the number of multipath reflections, α_l and τ_l respectively, the complex gain and delay associated with the lth multipath component. Assuming:

$$0 \leq \tau_1 \leq \tau_2 \leq \cdots \leq \tau_{L_m} \leq \tau_{max},$$

a channel is characterized as "flat" or "frequency selective" depending on the relation between τ_{max}, called the delay spread of the multipath channel, and T_c, the chip duration.

Narrowband Synchronous CDMA

When the chip duration T_c of a CDMA system far exceeds the delay spread, that is,

$$T_c \gg \tau_{max},$$

the CDMA system is considered "narrowband."[1] In this case, the channel response can be effectively modeled as:

$$f(t) = \sum_{l=1}^{L_m} \alpha_l \delta(t - \tau_l) \approx \alpha \delta(t)$$

and thus has a flat frequency response. The multipath channel introduces no distortion to the transmitted signal except a complex gain. Synchronous CDMA (S-CDMA) becomes possible. More specifically, with some base station coordination, the timing offsets of all uplink

---

[1]In the commercial world "narrowband CDMA" is often used to refer to the second-generation CDMA with data rate in the order of kilobits per second, whereas "wideband CDMA" means the third-generation PCS with data rate exceeding 1 Mbps.

CDMA Fundamentals

CDMA signals can be maintained to be within a small fraction of the chip duration T_c:

$$\tau_i/T_c \approx 0$$

The received signal, $y(t)$ in (1.2), reduces to:

$$y(t) = \sum_{k=-\infty}^{\infty} s_i(k) w_{i,k}(t - kT) \tag{1.4}$$

Recall that each signature waveform is a function of the spreading codes and the pulse-shaping function:

$$w_{i,k}(t) = \sum_{l=1}^{L} c_{i,k}(l) p(t - kT_c + T_c) \tag{1.5}$$

If $p(t)$ satisfies the Nyquist criterion, that is, $p(0) = 1$ and $p(kT) = 0$, $k \neq 0$, then with perfect timing one can sample $y(t)$ at the chip rate and stack L samples within a symbol period to obtain:

$$\mathbf{y}(k) = \begin{bmatrix} y(k,1) \\ y(k,2) \\ \vdots \\ y(k,L) \end{bmatrix} = \sum_{i=1}^{P} \begin{bmatrix} c_{i,k}(1) \\ c_{i,k}(2) \\ \vdots \\ c_{i,k}(L) \end{bmatrix} s_i(n) \tag{1.6}$$

where $y(k,l) = y((k-1)T_s + lT_c)$.

When orthogonal codes are employed, the ith signal can be recovered by simply despreading the sample vector with the spreading code:

$$\sum_{l=1}^{L} c_{i,k}(l) y(k,l) = \sum_{l=1}^{L} c_{i,k}^2(l) s_i(k) + \underbrace{\sum_{j \neq i} \sum_{l=1}^{L} c_{i,k}(l) c_{j,k}(l)\, s_j(k)}_{=0} = s_i(k)$$

The above operation, summarized in Figure 1.9, is equivalent to a matched filter (to the ith user's spreading waveform) followed by a symbol rate sampler. It can be easily shown that for S-CDMA, this simple operation eliminates all MAI and optimally recovers the original symbol sequence $\{s_i(k)\}$.

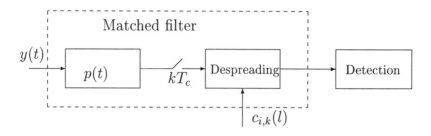

Figure 1.9 A synchronous CDMA receiver.

The Walsh codes (Hadamard) are the most commonly used orthogonal codes due to their modular structure that allows easy generation and fast spreading and despreading. Walsh codes of length 2^k can be recursively generated by the following rules:

$$\mathbf{H}_1 = \begin{bmatrix} 1 & 1 \\ 1 & -1 \end{bmatrix}$$

$$\mathbf{H}_k = \begin{bmatrix} \mathbf{H}_{k-1} & \mathbf{H}_{k-1} \\ \mathbf{H}_{k-1} & -\mathbf{H}_{k-1} \end{bmatrix}, \quad k = 2, 3, \cdots \quad (1.7)$$

The rows or columns of a $\mathbf{H}_k$ Hadamard matrix give a set of 2^k orthogonal codewords. The orthogonality comes from the fact that each pair of words has as many digit agreements as disagreements.

Frequency-Selective Channel and Wideband CDMA

A CDMA system is "wideband" when τ_{max} becomes comparable to T_c. In this case, the multipath channel introduces a convolutional effect to the modulating signature waveform $\{w_i(t)\}$. The received spreading waveform, referred to as the "effective signature waveform," is a function of the channel as follows:

$$\bar{w}_{i,k}(t) = w_{i,k}(t) * f_i(t) \quad (1.8)$$

Another way to represent $\bar{w}_{i,k}(t)$ is to express it as the convolution of the spreading codes, $c_{i,k}(l)$, and the "composite channel response,"

$$h_i(t) = p_t(t) * f_i(t) * p_t(t) \quad (1.9)$$

that accounts for the transmitter pulse-shaping filter, the actual multipath channel, delays, and the receiver filter:

$$\bar{w}_{i,k}(t) = \sum_{l=1}^{L} c_{i,k}(l) h_i(t - lT_c + T_c) \qquad (1.10)$$

Here (and for the rest of the book), I assume that the CDMA signal is already "coarsely synchronized" after the initial access. In this case, it is plausible to model $h_i(t)$ as FIR with support within the delay spread of the multipath channel: $[0 \quad L_c T_c]$. In most CDMA applications $L_c \ll L$, which means that intersymbol interference (ISI) is negligible in general.

The received signal at the base station becomes:

$$y(t) = \sum_{k=-\infty}^{\infty} s_i(k) \bar{w}_{i,k}(t - kT) \qquad (1.11)$$

Clearly, achieving orthogonality between channel-dependent effective signature waveforms is impossible. Each signal $s_i(k)$ in wideband CDMA is not only subject to MAI but also to the adverse interchip interference (ICI) effects that alter the spreading waveforms used at the transmitter.

Theoretically, if all effective signature waveforms are known to the receiver, optimum multiuser detection can be performed and the loss of orthogonality shall not lead to significant loss in CDMA performance. However, in real systems with implementational constraints (discrete-time sampling), the matched filter can only be "approximated" by reconstructing, in reverse order, each user's effective signature waveform as in Figure 1.10. Operations of multiuser detection hinge upon explicit knowledge of all users' codes, time delays, and attenuation.

Single-user receivers that rely solely on the desired user's channel information are still the dominating CDMA receivers used in today's wideband CDMA systems. The popular RAKE receiver attempts to enhance the output signal strength by combining the time-delayed versions of the signal from the desired user. The RAKE receiver can be easily realized with a set of correlators/despreaders, separated in time with fixed delays (e.g., by setting $\{\tau_i\}$ in Figure 1.10 to be half a chip apart), to detect the strongest multipath components. The outputs of each correlator/despreader are weighted and combined in the same way as a diversity combiner. With coarse synchronization, the RAKE receiver can operate without the exact timing information of the multipath components. In addition, the RAKE receiver is capable of tracking time-varying signals by aligning its middle correlator to the strongest multipath component. On the other hand, without accounting for the MAI, its performance is almost inferior (often significantly) to multiuser detectors. Chapters 2 and 4 provide more quantitative comparisons between the RAKE receiver and multiuser detectors.

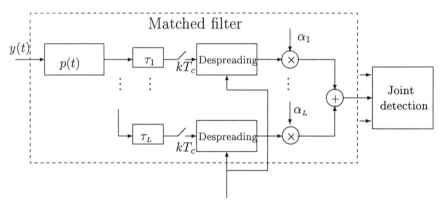

Figure 1.10 A broadband CDMA receiver.

In cases where multiuser detection is a critical imperative, the convolutional effect of the multipath channel poses several challenges to wideband CDMA communications:

1. The effective signature waveforms are channel-dependent and thus unknown to the receiver. Without such information, most multiuser detectors become inapplicable.

2. $\{\bar{w}_i(t)\}$ varies with the environment. The nonstationary component mandates the receiver to have certain tracking capabilities.

A majority of this book is devoted to techniques that cope with the adverse effects of multipath channels in wideband CDMA. In particular, I attempt to address the following question: Can high-performance

CDMA communication, similar to that offered by multiuser detection, be achieved in the presence of unknown multipath channels?

1.4 Spatial Diversity

In addition to the time-frequency resource, a CDMA-based wireless system is likely to employ multiple antennas to profit from the "space" resource through joint spatio-temporal-code domain processing. The basic idea of antenna array processing, illustrated in Figure 1.11, rests on the fact that each user in a wireless system has a uniquely associated spatial channel. Through these channels, the base station can perform spatially selective transmission/reception and therefore communicate with remote users in an efficient manner. In the following, we will establish the data model for CDMA with antenna diversities. More details on exploitation of the space domain will be discussed in later chapters.

The wireless channel model in (1.3) can be extended to incorporate spatial diversity induced by a receiver site antenna array.

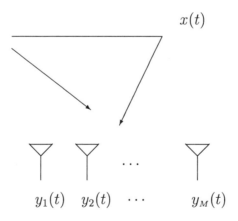

Figure 1.11 An antenna array base station.

Consider a radio signal impinging on an array of M antennas. Based on (1.3), the total channel response is the vector responses between individual antennas and the transmitter, and thus can be expressed as:

$$\mathbf{f}(t) = \begin{bmatrix} f_1(t) \\ f_2(t) \\ \vdots \\ f_M(t) \end{bmatrix} \quad (1.12)$$

where $f_m(t)$ denotes the channel response between the transmitter and the mth receiving antenna.

$\mathbf{f}(t)$ characterizes the vector channel of a single-input multi-output (SIMO) linear system. Similarly, when the delay spread is insignificant compared to the chip duration, $\mathbf{f}(t)$ becomes flat and reduces to:

$$\mathbf{f}(t) = \mathbf{a}\delta(t)$$

$\mathbf{a}$ is commonly referred to as the "spatial signature" of a flat channel. On the other hand, when multipath reflections are not coherent, i.e., they are different not only by a complex scalar, $\mathbf{f}(t)$ is called the "spatial filter." Mathematically, a spatial filter is simply an FIR vector filter.

Spatial channels introduce an additional dimension to the existing time-frequency resource as shown in Figure 1.5. Intelligent exploitation of the space dimension can lead to significant performance enhancement without extra spectrum allocation.

In the context of CDMA communications, the received signal from the antenna array is given by:

$$\begin{aligned}
\mathbf{y}(t) &= \sum_{i=1}^{P} \begin{bmatrix} f_{i,1}(t) \otimes x_i(t) \\ f_{i,2}(t) \otimes x_i(t) \\ \vdots \\ f_{i,M}(t) \otimes x_i(t) \end{bmatrix} = \sum_{i=1}^{P} \mathbf{f}_i(t) \otimes x_i(t) \quad (1.13) \\
&= \sum_{i=1}^{P} \sum_{k=-\infty}^{\infty} s_i(k) \left(\mathbf{f}_i(t) \otimes w_i(t-kT) \right) \\
&= \sum_{i=1}^{P} \sum_{k=-\infty}^{\infty} s_i(k) \bar{\mathbf{w}}_i(t-kT) \quad (1.14)
\end{aligned}$$

$\bar{\mathbf{w}}_i(t - kT)$ here denotes the 2D (space-time) effective signature waveform for the ith user.

Antenna arrays provide another dimension in which CDMA receivers can cancel out MAI and constructively combine SOI scattered in the time domain. For more details and explanations on this aspect, see Chapters 4 and 6.

1.5 Scope of This Book

For CDMA to reach its full potential, a number of technological issues, including the multipath channel effects, must be addressed. The aim of this book is to provide a signal processing perspective on some key CDMA problems and discuss their possible solutions. We realize that literature in this area is abundant. However, most of it is oriented toward high-performance detection, assuming that the effective spreading waveforms are known. This book includes an ensemble of research results and articles by the author that deal with wideband CDMA reception in the presence of "unknown" multipath channels and interfering users. Discussion will not be limited to any specific CDMA standard. Instead, the book's major theme deals with generic CDMA systems that can be described by (1.14). In particular, the following topics will be investigated in detail:

1. Blind reception in wideband CDMA with short codes: In order to take full advantage of the periodic structure for multiuser detection, one needs to be able to jointly detect CDMA signals characterized by the effective spreading waveforms. We will illustrate that this is possible in a frequency selective fading environment by deriving a subspace blind multiuser detection algorithm and a low complexity adaptive multiuser detector.

2. Space-time reception for wideband CDMA with long codes: For aperiodic CDMA where multiuser detection is problematic, the generic solution is to jointly exploit space and time diversities. The so-called 2D RAKE receiver is directly applicable to the most commonly used CDMA standard, IS-95. In addition to a full description of the 2D RAKE receiver structure, we will examine several possible implementation schemes of the 2D RAKE without requiring explicit knowledge of the multipath channels.

3. Downlink: Since the capacity of a typical CDMA system is uplink limited, most studies on CDMA focus on uplink communication. With the advances of multiuser detection techniques that can potentially lead to significant increases in uplink capacity, signal reception techniques in downlink CDMA become increasingly important. We will discuss several demodulation and detection strategies suitable for downlink CDMA. The behavior of different reception schemes in frequency selective fading channels will be investigated.

4. Multicarrier CDMA: The difficulties and complexity of uplink CDMA over a frequency selective fading channel are mainly due to the distortion of spreading waveforms. Recently, a new technique that combines the advantages of the simplicity of OFDM and the flexibility of CDMA was proposed. The aspects of MC-CDMA system design and receiver optimization will be discussed in this book.

5. Channel-aware medium access control (MAC): As both a modulation and multiple-access scheme, CDMA demonstrates many benefits of jointly designing the physical layer and the MAC layer. Problems that cannot be addressed using conventional methods may be solvable using cross-layer design principles. We will extend our discussion to the general MAC system and describe the notion of channel-aware MAC where a wireless system obtains substantial gains in performance by allowing information exchanges between the two layers. In particular, we will describe a packet radio network that exploits all forms of diversities (including the spatial diversity provided by antenna arrays) at both the physical and MAC layers.

References

[1] A. Samukic, "UMTS universal mobile telecommunications systems: Development of standards for the third generation," *IEEE Trans. Vehicular Technology*, **47**(4):1099–1104, November 1998.

[2] E. Dahlman, P. Beming, J. Knutsson, F. Ovesjo, M. Persson, and C. Roobol, "WCDMA – the radio interface for future mobile multimedia communications," *IEEE Trans. Vehicular Technology*, **47**(4):1104–1118, November 1998.

[3] M. K. Simon, J. K. Omura, R. A. Scholtz, and B. K. Levitt, *Spread Spectrum Communications Handbook*, revised edition, New York, NY: McGraw-Hill, 1994.

[4] A. J. Viterbi, *CDMA, Principles of Spread Spectrum Communication*, Reading, MA: Addison-Wesley, 1995.

[5] E. A. Lee and D. G. Messerschmitt, *Digital Communication*, New York: Kluwer-Academic, 1988.

[6] S. Verdú, "Optimum multiuser asymptotic efficiency," *IEEE Trans. on Communications*, **34**(9):890–897, September 1986.

[7] R. Lupas and S. Verdú, "Linear multiuser detectors for synchronous CDMA channels," *IEEE Trans. on Information Theory*, **35**(1):123–136, January 1989.

[8] Z. Xie, R. T. Short, and C. K. Rushforth, "A family of suboptimum detectors for coherent multiuser communications," *IEEE J. Selected Areas in Communications*, pages 683–690, May 1990.

[9] Z. Zvonar and D. Brady, "Suboptimum multiuser detector for synchronous CDMA frequency-selective Rayleigh fading channels," In *Globecom Mini-Conference on Communications Theory*, pages 82–86, 1992.

[10] R. Lupas and S. Verdú, "Near-far resistance of multiuser detectors in asynchronous channels," *IEEE Trans. on Communications*, **38**(4):496–508, April 1990.

[11] J. G. Proakis, *Digital Communications*, McGraw-Hill Book Company, Polytechnic Institute of New York, second edition, 1989.

[12] R. Price and P. E. Green, "A communication technique for multipath channels," *Proc. IRE*, **46**:555–570, March 1958.

[13] U. Mitra and H. V. Poor, "Adaptive receiver algorithm for

near-far resistant CDMA," *IEEE Trans. on Communications*, **43**(4):1713–1724, April 1995.

[14] A. Duel-Hallen, "Decorrelating decision-feedback multiuser detector for synchronous CDMA channel," *IEEE Trans. on Communications*, **41**(2):285–290, February 1993.

[15] S. L. Miller, "Training analysis of adaptive interference suppression for direct-sequence CDMA systems," *IEEE Trans. on Communications*, **44**(4):488–495, April 1996.

Chapter 2

CDMA With Short Codes: Direct Approaches

2.1 Introduction

Most CDMA modulators bear a structure similar to that depicted in Figure 1.3 and differ only in the manner in which they choose the spreading codes and timing. On the other hand, CDMA demodulators can be quite different depending on the multipath channel conditions and performance requirements. When the CDMA system is wideband, (1.8) indicates that the multipath induced interchip interference (ICI) can alter the spreading waveforms of transmitted signals. Without a proper receiver, the ICI may have a strong negative impact on the system capacity.

For CDMA with short codes, multiuser detection can be employed to enhance the reception performance by demodulating multiple signals at the same time. However, almost all multiuser detectors require explicit knowledge of users' spreading, timing, and, in the case of wideband CDMA, the frequency-selective channel coefficients. Because the system is time varying, this necessitates the use of training sequence or some computationally intensive channel parameter estimation algorithms. To save bandwidth and cost, there is an evident need for lower complexity receivers that can perform detection without explicit knowledge of the channel parameters.

An extreme choice would be the RAKE receiver, applicable to CDMA with any spreading codes, which coherently combines the signal-of-interest (SOI) scattered in the time domain. Despite its low complexity, we already point out in Chapter 1 that the conventional RAKE receiver relies exclusively on the statistics of the desired user and thus has fundamental difficulties in handling the MAI. For high-performance communication, it is necessary to find a better tradeoff between system cost and performance.

One technique that can potentially enhance the reception performance without introducing undue complexity is the constrained optimization approach commonly used in adaptive array processing. The central idea of constrained reception is to lock on the SOI while adaptively suppressing interference and noise using the minimum output energy (MOE) criterion. This concept was introduced to CDMA communications by Honig et al. under the assumption that the channels are frequency nonselective [1, 2].

In this chapter, we investigate the possibility of adaptive multiuser reception for quasi-synchronous CDMA communications over "unknown" frequency-selective fading channels. Toward this end, we combine the RAKE structure with the constrained optimization techniques and develop a self-adaptive receiver that has the advantages of the algorithms in [1–4] and overcomes their limitations. The new receiver is termed the decorrelating-RAKE (DRAKE) receiver for its ability to decorrelate multiple-access interference and combine desired signals stemming from the same sources [5]. Main features of the DRAKE receiver include (1) low complexity; (2) high performance, comparable to that of the adaptive MOE receivers for flat channels [1, 2]; and (3) completely blind in the sense that no knowledge of the desired user's channel response is required. Its extension to asynchronous and/or antenna array CDMA systems is straightforward.

The rest of the chapter is organized as follows. Section 2.2 presents some background on adaptive signal processing upon which the new CDMA receiver is developed. The DRAKE receiver is introduced in Section 2.3, and its performance is analyzed in Section 2.4. The efficacy of the new receiver is demonstrated with some simulation results.

2.2 Constrained Optimization

The key technique behind the blind adaptive receiver is the well-known constrained beamforming algorithm developed in antenna array applications [6]. To lay the groundwork, we shall first review the adaptive optimization algorithms and cast the CDMA problem into the same framework.

A well-studied problem in array signal processing is signal recovery in the presence of structured interference. The problem arises primarily in military antijamming communications and commercial wireless applications where the signal-of-interest is corrupted by intentional jamming signals and/or interference. Spatial beamforming constructively combines SOI and suppresses interference/noise by exploiting the spatial diversity.

To explain the idea mathematically, let:

$$\mathbf{y}(t) = \begin{bmatrix} y_1(t) \\ y_2(t) \\ \vdots \\ y_M(t) \end{bmatrix} = \mathbf{a}s(t) + \mathbf{i}(t) \qquad (2.1)$$

denote a vector of observations from an M-element antenna array; $s(t)$ is the signal-of-interest; $\mathbf{i}(t) = [i_1(t) \cdots i_M(t)]^T$ is the interference-plus-noise vector observed from the receivers; and $\mathbf{a}$ is the so-called spatial signature of the desired signal.

The spatial signature completely characterizes the spatial link between the user of interest and the antenna array. In particular, $\mathbf{a}$ can be explicitly decomposed into:

$$\mathbf{a} = \sum_{l=1}^{L_m} \alpha_l \mathbf{a}(\theta_l)$$

where L_m is the number of multipath reflections, α_l, θ_l, the complex gain and direction-of-arrival (DOA) of the lth multipath, respectively, and $\mathbf{a}(\theta_l)$ the array response vector determined uniquely by the DOA θ_l and the array configuration.

The signal recovery consists of applying a "weight vector" **g** to the observation vector with the intent of maximizing the output signal-to-interference-plus-noise ratio (SINR). Direct computation of the optimum weight vector requires complete knowledge of system spatial characteristics, including spatial signatures of the interfering signals. Often in practice, however, only the desired user's spatial signature **a** is available to the receiver antenna array.

Instead of maximizing the output SINR directly, the constrained optimization minimizes the mean-squared value of the weighted observations $E[|\mathbf{g}^H\mathbf{y}|^2]$ subject to a side constraint:

$$\mathbf{C}^H\mathbf{g} = \mathbf{e} \tag{2.2}$$

where **C** is the constraint matrix and **e** is a column vector of constraining values.

When $\mathbf{C} = \mathbf{a}$ and $\mathbf{e} = 1$, the optimization becomes a "constrained minimum output energy" (MOE) problem below:

$$\min_{\mathbf{g}} \mathbf{g}^H E\{\mathbf{y}\mathbf{y}^H\}\mathbf{g} = \min_{\mathbf{g}} \mathbf{g}^H \mathbf{R}_{\mathbf{yy}} \mathbf{g} \quad \text{subject to } \mathbf{a}^H\mathbf{g} = 1$$

In essence, the above receiver minimizes the total output energy while keeping the power of the SOI fixed. It can be shown that the solution to the above problem is identical to the minimum mean-squared error (MMSE) or maximum output SINR solution defined as:

$$\mathbf{g}_{\text{MMSE}} = \arg \min_{\mathbf{g}} E|\mathbf{g}^H\mathbf{y}(t) - s(t)|^2$$

Indeed, if the output power of the SOI is fixed while the total output energy is being minimized, the output SINR is clearly maximum, leading to the MMSE solution. Unlike direct implementation of MMSE reception, operating the constrained MOE requires only knowledge of **a**, the spatial signature of the SOI. No information on the interfering signals is required.

It is shown in that using Lagrange multipliers, the solution $\mathbf{g}_o$ for the general constrained adaptive array processor is given by:

$$\mathbf{g}_o = \mathbf{R}_{\mathbf{yy}}^{-1}\mathbf{C}\left(\mathbf{C}^H\mathbf{R}_{\mathbf{yy}}^{-1}\mathbf{C}\right)^{-1}\mathbf{e}$$

Further, the solution to the constrained adaptive beamformer can be found iteratively by decomposing $\mathbf{g}_o$ into two orthogonal components (Figure 2.1) — a completely nonadaptive component within the span of $\mathbf{C}$ to explicitly enforce the constraints [6]:

$$\begin{aligned} \mathbf{g}_c &= \mathbf{P_C}\, \mathbf{g}_o \\ &= \mathbf{C}\left(\mathbf{C}^H\mathbf{C}\right)^{-1}\mathbf{C}^H\mathbf{R}_y^{-1}\mathbf{C}\left(\mathbf{C}^H\mathbf{R}_y^{-1}\mathbf{C}\right)^{-1}\mathbf{e} \\ &= \mathbf{C}\left(\mathbf{C}^H\mathbf{C}\right)^{-1}\mathbf{e} \end{aligned}$$

and the adaptive component, $\mathbf{g}_a$, that is orthogonal to $\mathbf{g}_c$ and adaptable to mitigate the interference. The receiver in Figure 2.1 can be implemented using the standard least mean squares (LMS) algorithm of complexity $O(M)$ in each step [7].

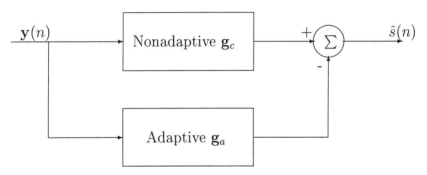

Figure 2.1 Adaptive constrained optimization.

CDMA Analogy

To cast the CDMA problem into the antenna array framework, let us first examine (1.10) in Chapter 1, where the effective signature waveform in wideband CDMA is expressed as the convolution of the spreading codes and the FIR channel response. Because of the channel effect, the duration of $\{\bar{w}_i\}$ will slightly exceed the symbol period, leading to minor ISI. As for most CDMA receivers, we assume that each uplink signal is already coarsely synchronized within the maximum support of the composite channel, $L_c T_c$. In other words, the base

30   Signal Processing Applications in CDMA Communications

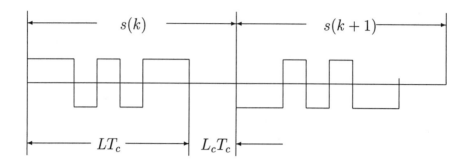

Figure 2.2 ISI-free CDMA communications.

station knows roughly the starting point of each signal. To avoid ISI, we further assume a guard period of length $L_c T_c$ is used so that all effective signature waveforms are limited to the current symbol period. See Figure 2.2 for illustration.

Mathematically, adding the guard period is equivalent to pending zeros to the spreading codes. Note that the symbol duration is now $L + L_c$ chips. Under these assumptions, the ISI-free chip-rate CDMA signals within a symbol period can be expressed in a vector as follows,

$$\mathbf{y}(k) = \begin{bmatrix} y_1(k) \\ \vdots \\ y_{L+L_c}(k) \end{bmatrix} = \sum_{i=1}^{P} \bar{\mathbf{w}}_i s_i(k) \qquad (2.3)$$

Rewriting the discrete-time version of (1.10) in a vector form, the $(L + L_c) \times 1$ effective signature waveform can be expressed as:

$$\bar{\mathbf{w}}_i = \begin{bmatrix} c_i(1) & \cdots & 0 \\ \vdots & \ddots & \vdots \\ c_i(L) & \cdots & c_i(1) \\ \vdots & \ddots & \vdots \\ 0 & \cdots & c_i(L) \end{bmatrix} \begin{bmatrix} h_i(L_c) \\ \vdots \\ h_i(1) \end{bmatrix}$$

$$\overset{\text{def}}{=} [\mathbf{c}_{iL_c} \cdots \mathbf{c}_{i1}] \mathbf{h}_i \overset{\text{def}}{=} \mathbf{C}_i \mathbf{h}_i \qquad (2.4)$$

Each $\mathbf{c}_{il}$ represents a delayed version of the spreading codes, and $\mathbf{C}_i$ is the Hankel code matrix that defines the span of the effective signature waveform.

The above equation suggests that the signature waveform of the SOI, $\bar{\mathbf{w}}$, is uniquely determined by the "known" code matrix, $\mathbf{C}$, and the unknown channel vector, $\mathbf{h}$. This observation reduces the number of unknowns in $\bar{\mathbf{w}}_i$ from $L + L_c$ to L_c, the number of channel coefficients.

Without loss of generality, we let $s_1(n)$ be the signal of interest and rewrite the data vector (2.3) as:

$$\mathbf{y}(k) = \bar{\mathbf{w}}_1 s_1(k) + \mathbf{u}(k)$$

Here $\mathbf{u}(n) = \sum_{i=2}^{P} \bar{\mathbf{w}}_i s_i(n) + \mathbf{v}(n)$ denotes the MAI plus noise. For simplicity, we shall drop the subscript and consider the recovery of $s(k)$ from observations:

$$\mathbf{y}(k) = \bar{\mathbf{w}} s(k) + \mathbf{u}(k) \tag{2.5}$$

The similarity between the antenna output vector (2.1) and CDMA signal vector in (2.5) is evident. Table 2.1 highlights the counterparts between the two applications. The structural information in CDMA is actually richer than that in an array system. In particular, the CDMA receiver has full knowledge of the delayed spreading vectors $\{\mathbf{c}_i\}$ and the maximum delay spread L_c in the composition of $\bar{\mathbf{w}}$, whereas in an array system the DOAs $\{\theta_l\}$ and the number of multipath reflections L_a must be estimated [8, 9]. The only unknowns in CDMA are the channel coefficients $\{h(l)\}$. In the context of an antenna array system, this is equivalent to the situation in Figure 2.3, where all DOAs are perfectly known to the receivers.

Clearly, the constrained MOE beamforming techniques are directly applicable to CDMA. Honig et al. were the first to apply such techniques to the CDMA reception problem. When the multipath channels are flat, the effective signature waveform $\bar{\mathbf{w}} = \mathbf{w}$, and thus is perfectly known to the receiver. It is suggested to use the constrained MOE technique to adaptively suppress interference at the receiver. The method offers significant performance gain without knowledge of the interfering users' signature waveforms or timing information. On the

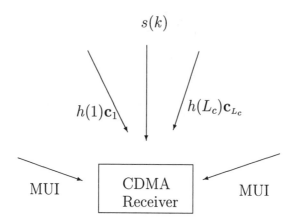

Figure 2.3 Resemblance between CDMA and antenna arrays.

other hand, when the channel is frequency-selective, one may still have to use training sequences to obtain knowledge of the effective signature waveform $\bar{\mathbf{w}}$. In this case, direct application of the constrained MOE receiver becomes less attractive.

In the remainder of this book, I shall refer to the constrained MOE method in [1] that requires explicit knowledge of the desired user's effective signature waveform as the MOE receiver.

Antenna Array	CDMA
Antenna elements (M)	# of chips per symbol ($L + L_c$)
Spatial signature $\mathbf{a}$	Effective signature waveform $\bar{\mathbf{w}}$
Array response vector $\{\mathbf{a}(\theta_l)\}$	Delayed spreading vectors $\{\mathbf{c}_i\}$
Complex gains $\{\alpha_l\}$	Channel coefficients $\{h_i\}$

Table 2.1 Counterparts in CDMA and antenna array systems.

2.3 Decorrelating RAKE Receivers

In wideband CDMA where users' effective signature waveforms are unknown to the base station, new constrained adaptive receivers have

CDMA With Short Codes: Direct Approaches

to be developed. Fortunately, unlike the antenna array system where the spatial signature vector is totally unknown to the receivers, there is abundant information about the signature waveforms in CDMA. As explained earlier, $\bar{\mathbf{w}}$ can be decomposed into $\mathbf{Ch}$ where $\mathbf{C}$ is known a priori. In the antenna array context, this is analogous to the situation where each direction-of-arrival (DOA) of the desired user is known within a complex scalar. Using knowledge of these DOAs, one can in principle extract the desired signal from each direction defined by $\mathbf{c}_l$ without causing signal cancellation. This is exactly the idea behind the blind receiver to be developed in ensuing sections.

2.3.1 Constrained MOE

Rewrite (2.5) as follows:

$$\begin{aligned}\mathbf{y}(n) &= \bar{\mathbf{w}}s(n) + \mathbf{u}(n) = \mathbf{Ch}s(n) + \mathbf{u}(n) \\ &= [\mathbf{c}_1 \cdots \mathbf{c}_{L_c}] \begin{bmatrix} h(1) \\ \vdots \\ h(L_c) \end{bmatrix} s(n) + \mathbf{u}(n)\end{aligned} \quad (2.6)$$

Our approach of retrieving $s(n)$ from $\mathbf{y}(n)$ is to utilize all signal energy and, at the same time, suppress the MAI and noise through constrained MOE filtering. The goal here is quite different from that of the conventional RAKE receiver, in which the MAI cannot be cancelled because of the fixed spreading operation. As will be shown in the ensuing discussion, the new receiver replaces the despreader with a more flexible filter and is capable of eliminating all MAI.

From (2.5), it is seen that the desired signal is a linear combination of signals projected onto a set of delayed code vectors, $\{\mathbf{c}_l\}_{l=1}^{L_c}$. Taking advantage of this structure, we will show below that the self-adaptive stage-2 receiver depicted in Figure 2.4 can provide a near-optimum estimate of $s(n)$ directly from $\mathbf{y}(n)$ with low cost.

The new receiver has two stages, with stage-1 comprised of a set of adaptive weight vectors $\{\mathbf{g}_l\}$ and stage-2 a coherent combiner. The idea is the following. Since only the channel coefficients $\{h(l)\}_{l=1}^{L_c}$ are unknown, we can first construct a special set of weight vectors to extract the desired signal along "individual" code vectors while

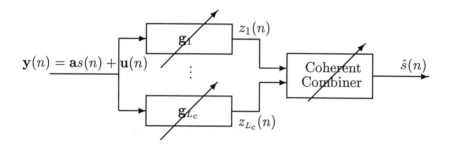

Figure 2.4 The decorrelating-RAKE receiver.

suppressing all MAI. After that, we can then constructively combine the extracted signals to achieve near-optimum signal estimation. Note that without isolating the SOI along each delayed code vector, the MOE type of receiver will experience signal cancellation, leading to significant performance degradation.

The extracting operation in stage-1 can be mathematically represented as:

$$\mathbf{C}^H \mathbf{g}_l = [0 \cdots 1 \cdots 0]^T \stackrel{\text{def}}{=} \mathbf{1}_l, \quad l = 1, \cdots, L_h$$

where $\mathbf{1}_l$ is a vector with all elements 0s except 1 at the lth position. Referring to Figure 2.4, the output of the lth filter (or the lth arm) is given by:

$$z_l(n) = \mathbf{g}_l^H \mathbf{y}(n) = \mathbf{g}_l^H \bar{\mathbf{w}} s(n) + \mathbf{g}_l^H \mathbf{u}(n) = \underbrace{\mathbf{g}_l^H \mathbf{C}}_{\mathbf{1}_l^H} \mathbf{h} s(n) + \mathbf{g}_l^H \mathbf{u}(n)$$
$$= h(l)s(n) + \mathbf{g}_l^H \mathbf{u}(n) = h(l)s(n) + e_l(n) \quad (2.7)$$

where $e_l(n)$ denotes the effective noise and interference after filtering. Clearly, $\mathbf{g}_l$ picks up the desired signal along $\mathbf{c}_l$ while avoiding signal cancellation along $\mathbf{c}_1, \cdots, \mathbf{c}_{l-1}, \mathbf{c}_{l+1}, \cdots, \mathbf{c}_{L_h}$. The use of L_c weight vectors allow us to extract all signals at different delays.

With a total of L_c arms, the signal power after the first stage is $\sum_{l=1}^{L_c} |h(l)|^2 = \|\mathbf{h}\|^2$. To obtain the best signal estimate, one should maximize the signal-to-interference-plus-noise ratio (SINR) at the output of each arm defined as:

$$\text{SINR}_l = \frac{|h(l)|^2 E[s(n)s^*(n)]}{E[e_l(n)e_l^*(n)]} = \frac{|h(l)|^2}{\mathbf{g}_l^H \mathbf{R}_{\mathbf{uu}} \mathbf{g}_l}$$

Note the output signal power is fixed at $|h(l)|^2$, maximizing SINR_l is equivalent to minimizing the output power $E[z_l(n)z_l^*(n)]$. Therefore, the constrained MOE receiver is readily applied:

$$\mathbf{g}_l = \arg\min_{\mathbf{g}_l} \mathbf{g}_l^H \mathbf{R}_{\mathbf{yy}} \mathbf{g}_l \quad \text{subject to} \quad \mathbf{C}^H \mathbf{g}_l = \mathbf{1}_l \qquad (2.8)$$

The adaptive rules will be outlined in the next section.

It has been shown that the batch-mode MOE receiver is equivalent to an MMSE receiver under the constraint of $\mathbf{g}^H \bar{\mathbf{w}} = 1$. Since $\bar{\mathbf{w}}$ is unknown, each arm in the new receiver only provides a constrained MOE estimate of the signal. Nevertheless, $\mathbf{g}_l$ is capable of eliminating all interference and perfectly recovering the signal in the absence of noise [3].

Stacking the filter outputs from all arms in a more compact vector form:

$$\mathbf{z}(k) \overset{\text{def}}{=} \begin{bmatrix} z_1(k) \\ \vdots \\ z_{L_c}(k) \end{bmatrix} = \mathbf{h}s(k) + \begin{bmatrix} e_1(k) \\ \vdots \\ e_{L_c}(k) \end{bmatrix} \overset{\text{def}}{=} \mathbf{h}s(k) + \mathbf{e}(k) \qquad (2.9)$$

The second stage of the proposed receiver coherently combines the outputs from all arms to further enhance the SINR. In principle, construction of the optimum combining vector, $\mathbf{R}_{\mathbf{xx}}^{-1}\mathbf{h}$, requires explicit knowledge of the channel coefficients. However, because of the interference decorrelation and noise suppression in the first stage, the total signal power in $\mathbf{z}(n)$ becomes significantly higher than that of $\mathbf{e}(n)$. In this case, $\mathbf{R}_{\mathbf{zz}} = \mathbf{h}\mathbf{h}^H + \mathbf{R}_{\mathbf{ee}} \approx \mathbf{h}\mathbf{h}^H$. This enables us to approximate the optimum combining vector using the principal eigenvector of $\mathbf{R}_{\mathbf{zz}}$, which can be obtained directly using standard decomposition techniques [10].

The approximation here is a common trade-off between optimality and complexity. The same technique is used in practical RAKE receivers. Note in an interference-limited CDMA environment, the performance loss due to suboptimum combining of the present receiver should be far less than that of the conventional RAKE receiver.

In summary, the proposed blind receiver restores the desired user's signal from the CDMA output in two steps:

- Step 1: Extract the desired signal along each of the delayed code vectors using the constrained adaptive MOE receivers (2.8).

- Step 2: Coherently combine the first stage filter outputs to further enhance the SINR of the final signal estimate.

Step 2 is critical in mobile communications. By doing so, we overcome the major difficulty of the algorithm proposed in [4], which is essentially a single-arm receiver relying solely on $\mathbf{g}_1$. Due to channel variations, performance of the single-arm receiver is susceptible to shifts of the dominant multipath component and thus lacks the robustness for practical applications. The scheme described here constructively combines signals from all L_c arms, thereby offering strong resistance against fading and timing ambiguity. The "worst case" output SINR, which usually determines the capacity of a wireless system, is greatly improved. Relative to the MOE approach that requires explicit knowledge of $\bar{\mathbf{w}}$ [1, 2], the blind method here has slightly higher complexity, depending on the maximum spread of the wireless channel. The fact that a long-delay multipath reflection is substantially weaker than a short-delay signal permits the use of a limited number of arms in the receiver without sacrificing performance.

It is worth pointing out that by replacing $\mathbf{g}_l$ with a despreader, $\mathbf{c}_l$, the receiver in Figure 2.4 reduces to a conventional RAKE receiver. Although resembling it in structure, the proposed method is fundamentally different from the well-known RAKE receiver that coherently combines desired signals at multiple fingers. The major difference is the fact that the new receiver is decorrelating in nature. After converging to its optimum values, each arm serves as a decorrelating receiver in the absence of noise. In other words,

$$\mathbf{g}_{l,\text{opt}}^H [\bar{\mathbf{w}}_1 \ \bar{\mathbf{w}}_2 \ \cdots \ \bar{\mathbf{w}}_P] = [h_1(l) \ 0 \cdots 0]$$

provided that $P + L - 1 \leq L + L_c$. Then $z_l(k) = \mathbf{g}_l^H \mathbf{y}(k) = h(l)s(k)$, leading to perfect signal recovery. This is certainly not possible for a single-user RAKE receiver. In the presence of noise, each arm is a constrained MOE receiver and thus provides performance far superior to a regular RAKE receiver. To differentiate, we shall refer to the new receiver as the decorrelating-RAKE (DRAKE) receiver.

2.3.2 Adaptive Implementation

Both the weight vectors in the first stage and the coherent combining coefficients in the second stage need to be estimated from the received signals. When utilized in a non-stationary environment, it is necessary to implement the DRAKE receiver adaptively in order to track the time-varying channels.

The constrained receiver vectors, $\{\mathbf{g}_l\}_{l=1}^{L_c}$, are obtained via standard LMS algorithms. The adaptive rules are briefly reviewed below; readers can referred to [6, 7, 11] for more discussion.

Denote the output power of the lth arm as:

$$J_{\text{MOE}} = \mathbf{g}_l^H \mathbf{R}_{\mathbf{yy}} \mathbf{g}_l$$

The objective herein is to adaptively search for the receiver vector, $\mathbf{g}_l$, that minimizes J_{MOE} subject to $\mathbf{C}^H \mathbf{g}_l = \mathbf{1}_l$.

The gradient of the cost function is given by:

$$\nabla_{\mathbf{g}_l}(J_{\text{MOE}}) = 2\mathbf{R}_{\mathbf{yy}} \mathbf{g}_l, \quad l = 1, \cdots, L_h \quad (2.10)$$

Approximating the autocorrelation matrix by the outer product of the instantaneous received vector $\mathbf{y}(k)$ yields:

$$\nabla_{\mathbf{g}_l}(J_{\text{MOE}}) \approx 2\mathbf{y}\mathbf{y}^H \mathbf{g}_l, \quad l = 1, \cdots, L_h \quad (2.11)$$

In order to restrict our search direction in the constrained subspace, we need to find the projection of the gradient of the output energy onto the subspace orthogonal to $\mathbf{C}$. Upon defining the orthogonal projection matrix $\mathbf{P}_{\mathbf{C}}^{\perp} = \mathbf{I} - \mathbf{C}^H(\mathbf{CC}^H)^{-1}\mathbf{C}$, we arrive at the following recursive rule:

$$\begin{aligned} \mathbf{g}_l(k+1) &= \mathbf{g}_l(k) - \mu \mathbf{P}_{\mathbf{C}}^{\perp} \mathbf{y}(k) \mathbf{y}^H(k) \mathbf{g}_l(k) \\ &= (\mathbf{I} - \mu \mathbf{P}_{\mathbf{C}}^{\perp} \mathbf{y}(k) \mathbf{y}^H(k)) \mathbf{g}_l(k) \end{aligned} \quad (2.12)$$

It is easy to show if we use $\mathbf{g}_l(0) = \mathbf{C}^\dagger \mathbf{1}_l$ (where $\mathbf{C}^\dagger$ denotes the pseudoinverse of $\mathbf{C}$) as the initial weight vector, the constraint is satisfied in each iteration. The choice of step-size μ represents a compromise between rate of convergence and steady-state excess error.

Both standard and fast adaptive eigen-decomposition techniques [12] can be employed to estimate the principal vector of $\mathbf{R}_{\mathbf{zz}}$ for coherent combining in the second stage. Note since $L_c \ll L$, the computational cost (at most $O(L_c^3)$) at the second stage is negligible relative to that of the first stage.

2.4 Performance Analysis

Several issues regarding the performance of the DRAKE receiver are addressed in this section. While misadjustment and convergence are among the most important ones in adaptive algorithms, it is theoretically interesting to investigate the performance limit of the DRAKE receiver and compare it with that of the true MMSE receiver. In this section, we first evaluate the optimality of the DRAKE receiver using the output SINR as a performance measure. The steady-state behavior of the blind receiver is then studied. The results will reveal the efficacy of the new method and provide important insight into system implementation.

2.4.1 Receiver Optimality

The analysis here only provides a performance bound for batch-mode implementation of the DRAKE algorithm. I showed earlier that asymptotically the batch-mode MOE receiver converges to the MMSE receiver. The exact performance depends on the distribution of the interfering users' power, channel characteristics, and noise statistics. For simplicity, assume that the estimation errors in the first step dominate the errors in the second step, which allows one to ignore the overall interference-plus-noise correlation in the process of combining.

The closed-form optimum weight vector for the lth arm can be

CDMA With Short Codes: Direct Approaches

obtained via Lagrange multipliers [9]:

$$\mathbf{g}_{l,\text{opt}} = \mathbf{R}_{\mathbf{yy}}^{-1}\mathbf{C}^H(\mathbf{C}\mathbf{R}_{\mathbf{yy}}^{-1}\mathbf{C}^H)^{-1}\mathbf{1}_l \qquad (2.13)$$

It is not hard to show that in the batch mode, the weight vector in each arm will converge to its optimum value with probability 1, as the number of data observations approaches infinity. In the DRAKE receiver, outputs from multiple arms are coherently combined. The optimum output of the DRAKE receiver is thus given by:

$$\sum_{l=1}^{L} h(l)^* \mathbf{g}_{l,\text{opt}}^H \mathbf{y}(n)$$

In effect, one can model the stage-2 receiver using one receiving weight vector below:

$$\mathbf{g}_{\text{DRAKE}} = \sum_{l=1}^{L_h} h(l)\mathbf{g}_{l,\text{opt}} = \mathbf{R}_{\mathbf{yy}}^{-1}\mathbf{C}(\mathbf{C}^H\mathbf{R}_{\mathbf{yy}}^{-1}\mathbf{C})^{-1}\mathbf{h} \qquad (2.14)$$

Therefore, the maximum SINR that can be achieved by the DRAKE receiver is given by:

$$\begin{aligned}\text{SINR}_{\text{DRAKE}} &= \frac{E[\mathbf{g}_{\text{DRAKE}}^H \bar{\mathbf{w}} s(n) s^*(n) \bar{\mathbf{w}}^H \mathbf{g}_{\text{DRAKE}}]}{E[\mathbf{g}_{\text{DRAKE}}^H \mathbf{u}\mathbf{u}^H \mathbf{g}_{\text{DRAKE}}]} \\ &= \frac{\|\mathbf{h}\|^4}{\mathbf{g}_{\text{DRAKE}}^H \mathbf{R}_{\mathbf{uu}} \mathbf{g}_{\text{DRAKE}}}\end{aligned} \qquad (2.15)$$

Similarly, it can be derived that the output SINR corresponding to the MMSE receiver, $\mathbf{g}_{\text{MMSE}} = \mathbf{R}_{\mathbf{yy}}^{-1}\bar{\mathbf{w}} = \mathbf{R}_{\mathbf{yy}}^{-1}\mathbf{Ch}$, is given by:

$$\text{SINR}_{\text{MMSE}} = \frac{|\mathbf{g}_{\text{MMSE}}\bar{\mathbf{w}}|^2}{\mathbf{g}_{\text{MMSE}}^H \mathbf{R}_{\mathbf{uu}} \mathbf{g}_{\text{MMSE}}} \qquad (2.16)$$

whereas the maximum output SINR corresponding to the single-arm receiver, $\mathbf{g}_{1,\text{opt}}$, is:

$$\text{SINR}_1 = \frac{|h(1)|^2}{\mathbf{g}_{1,\text{opt}} \mathbf{R}_{\mathbf{uu}} \mathbf{g}_{1,\text{opt}}} \qquad (2.17)$$

The SINRs in (2.15), (2.16), and (2.17) provide the MSE upper bounds for the DRAKE receiver, the adaptive MMSE receiver, and the single-arm receiver, respectively.

Proposition 1 *When the system SNR is high,*

$$\text{SINR}_{\text{MMSE}} \geq \text{SINR}_{\text{DRAKE}} \geq \text{SINR}_1$$

In the noise-free cases, all three receivers are zero-forcing after converging to their corresponding optimum values and thus yield the same asymptotic performance. In the presence of noise, both $\text{SINR}_{\text{DRAKE}}$ and SINR_1 are clearly upper-bounded by $\text{SINR}_{\text{MMSE}}$. On the other hand, since the principal vector of $\mathbf{R}_{\mathbf{zz}}$ is the optimum combining vector when the system SNR is high, $\text{SINR}_{\text{DRAKE}}$ must be higher than SINR_1. The gaps between $\text{SINR}_{\text{DRAKE}}$ and $\text{SINR}_{\text{MMSE}}$, and SINR_1 and $\text{SINR}_{\text{DRAKE}}$, are determined by the angles between the weight vectors.

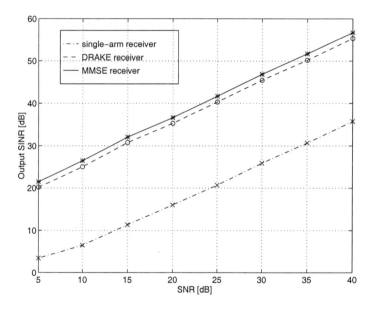

Figure 2.5 SINR bound for three types of receivers.

Since the single-arm receiver only uses signal output from one arm and discards the others, unless there are no multipath components and the system is perfectly synchronized (i.e., $\mathbf{h} = [h(1), 0 \cdots 0]^T$), the performance of the single-arm receiver is always inferior to that of the DRAKE receiver. The degradation in performance can be significant when the wireless system is dynamic.

For illustration, we simulate a 10-user CDMA system over fading channels and numerically calculate theoretic SINRs for the three approaches. The results are illustrated in Figure 2.5 under different input SNRs. As shown, the DRAKE receiver significantly outperforms the single-arm receiver. In the meantime, the performance gap between the MMSE receiver and the DRAKE receiver is very small.

2.4.2 Steady-State Behavior

In adaptive filtering, we are more interested in the steady-state behavior and especially the excess MSE of the adaptive algorithm. To derive the excess MSE of the DRAKE receiver, we first use the results in [1] to analyze the performance of the receiving filter at each arm. The effect of coherent combining on the final steady-state SINR is then derived. Our analysis here follows standard steps, which is common in the study of adaptive LMS algorithms.

For the lth arm, define $\Delta \mathbf{g}_l(n) = \mathbf{g}_l(k) - \mathbf{g}_{l,\mathrm{opt}}$, and subtract $\mathbf{g}_{l,\mathrm{opt}}$ from both sides of (2.12),

$$\Delta \mathbf{g}_l(k+1) = (\mathbf{I} - \mu \mathbf{P}_{\mathbf{C}}^{\perp} \mathbf{y}(k) \mathbf{y}^H(k)) \Delta \mathbf{g}_l(k) - \mu \mathbf{P}_{\mathbf{C}}^{\perp} \mathbf{y}(k) \mathbf{y}^H(k) \mathbf{g}_{l,\mathrm{opt}}$$

Following the steps used in [1] and noticing that the same adaptation rules are used here except that their projection vector is now the orthogonal projection matrix $\mathbf{P}_{\mathbf{C}}^{\perp}$, we obtain the excess MSE expression as:

$$J_{l,\mathrm{ex}} \approx \xi_{l,\mathrm{min}} \frac{\frac{\mu}{2} tr(\mathbf{P}_{\mathbf{C}}^{\perp} \mathbf{R}_{\mathbf{yy}})}{1 - \frac{\mu}{2} tr(\mathbf{P}_{\mathbf{C}}^{\perp} \mathbf{R}_{\mathbf{yy}})} \qquad (2.18)$$

where $\xi_{l,\mathrm{min}}$ is the minimum output energy given by:

$$\xi_{l,\mathrm{min}} = \mathbf{g}_{l,\mathrm{opt}}^H \mathbf{R}_{\mathbf{yy}} \mathbf{g}_{l,\mathrm{opt}} = \mathbf{1}_l^H \left(\mathbf{C}^H \mathbf{R}_{\mathbf{yy}}^{-1} \mathbf{C} \right)^{-1} \mathbf{1}_l \qquad (2.19)$$

Because the output signal power is fixed at $|h(l)|^2$, the constrained MOE at the lth arm output is given by:

$$J_{l,\mathrm{min}} = \xi_{l,\mathrm{min}} - |h(l)|^2 \qquad (2.20)$$

Therefore, the steady-state MSE and the steady-state output SINR, respectively, are:

$$\text{MSE}_{\text{ss},l} = J_{l,\min} + J_{l,\text{ex}}, \quad \text{SINR}_{\text{ss},l} = \frac{|h(l)|^2}{J_{l,\min} + J_{l,\text{ex}}} \qquad (2.21)$$

To obtain the steady-state MSE of the DRAKE receiver output, we again assume that the power of $\mathbf{e}(n)$ in $\mathbf{z}(n)$ is small relative to the signal strength, and the combining vector can be approximated by $\mathbf{h}$.

Rewrite the output from the lth arm as:

$$z_l(k) = h(l)s_1(k) + e_l(k), \quad l = 1, \cdots, L_c \qquad (2.22)$$

Here $e_l(n)$ denotes the overall error and $E[e_l(\infty)e_l^*(\infty)] = J_{l,\min} + J_{l,\text{ex}}$. The output of the coherent combiner is given by:

$$\hat{s}(k) = \sum_{l=1}^{L_c} h^2(l)s(k) + \sum_{l=1}^{L_c} h(l)e_l(k) = \|\mathbf{h}\|^2 s(k) + \sum_{l=1}^{L_c} h(l)e_l(k)$$

For analytical tractability, we further assume that the excess errors at different arms are independent[1]; then the final steady state SINR is given by:

$$\text{SINR}_{\text{ss,DRAKE}} = \frac{\|\mathbf{h}\|^4}{\sum_{l=1}^{L_c} |h(l)|^2 (J_{l,\min} + J_{l,\text{ex}})} \qquad (2.23)$$

where $J_{l,\text{ex}}$ and $J_{l,\min}$ are given in (2.18) and (2.20), respectively.

Although several seemingly strong assumptions are invoked in the above derivation, the final result in (2.23) matched well with our simulation results under different setups. The expression of the steady-state SINR provides an important guideline in the design and implementation of the DRAKE receiver.

2.5 Examples

We now present some numerical results to demonstrate the performance of the DRAKE receiver. In all examples, the channel responses

---
[1]This is certainly not valid in general but is frequently used in the performance analysis of adaptive filtering.

are generated using the well-known multiray model. The pulse function is raised-cosine with a roll-off factor of 0.5. For each user, the multipath delay and the number of multipath components are assumed to be uniformly distributed within $[0 \quad 3T_c]$ and $[1, 10]$, respectively. The number of first-stage weight vectors in the proposed DRAKE receiver is thus fixed at 3. The spreading factor is set to be 32 and power control within 3 dB is assumed.

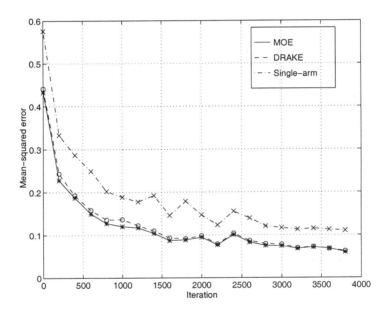

Figure 2.6 DRAKE vs. MOE and single-arm receivers.

Example 1: The first case involves 10 almost-synchronized CDMA users. The performance of three types of receivers, namely, the blind DRAKE receiver, the single-arm receiver [3], and the MOE [1, 2], is compared and the results are illustrated in Figure 2.6. The MSEs of the DRAKE receiver outputs are nearly identical to that of the MOE receiver, whereas the MSEs of a single-arm receiver are consistently higher. It is also observed that all three approaches converge rapidly at almost the same rate. After 3000 iterations, all three receivers reached the steady-state, and the gap between the performance of the DRAKE receiver and the MOE receiver almost vanished.

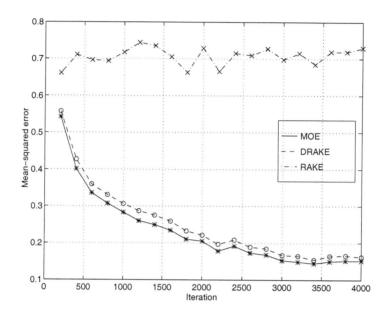

Figure 2.7 DRAKE vs. MOE and conventional RAKE receivers.

Example 2: Figure 2.7 compares the performance of the conventional RAKE receiver to that of the MOE and the DRAKE receivers in a 15-user setup. Not surprisingly, both multiuser receivers offer significantly better signal estimates than the conventional RAKE receiver. Further, the performance of the blind DRAKE receiver is consistently close to that of the MOE receiver.

Example 3: Under 3 dB SNR, we examine the excess MSE of a single-arm output for a particular user and gradually increased the number of total users in the system. As shown in Figure 2.8, where the solid line and "o"s represent the theoretic predictions whereas the dotted line and "x"s represent the simulation results, the two curves match extremely well when the total number of users is under 10. As the number of users increases, the theoretic values tend to overpredict the excess MSE.

Example 4: In the last example, we compare the near-far resistance of the blind DRAKE receiver with the near-far resistant MOE receiver.

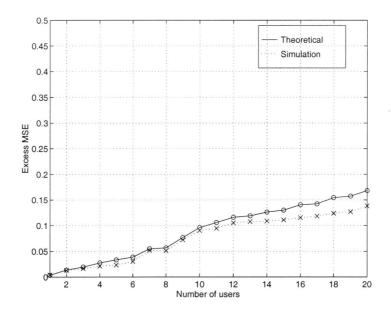

Figure 2.8 Excess MSEs.

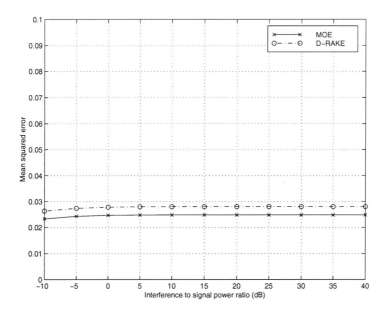

Figure 2.9 Near-far resistance: DRAKE vs. MOE.

A 10-user CDMA system is simulated. We fix the signal strength for the desired user and adjust the power level of all the other interfering users. The SNR is set to be 3 dB. Figure 2.9 plots the theoretical output MSEs of the two receivers for a range of interference to signal power ratios. It can be seen that the DRAKE receiver and the MOE receiver possess similar near-far resistance properties.

2.6 Conclusion

A decorrelating-RAKE (DRAKE) receiver for CDMA communications over frequency-selective fading channels has been developed in this chapter. The new receiver presents a good tradeoff between the low-complexity/low-performance RAKE receiver and high-cost/high-performance blind multiuser detectors described in Chapter 3. The asymptotic optimality and the steady-state behaviors of the DRAKE algorithm have been investigated. The results show that the new receiver is an ideal candidate for WCDMA in a mobile environment.

References

[1] M. L. Honig, U. Madhow, and S. Verdú, "Blind adaptive multiuser detection," *IEEE Trans. on Information Theory*, **41**(4):944–96, July 1995.

[2] J. B. Schodorf and D. B. Williams, "A constrained adaptive diversity combiner for interference suppression in CDMA systems," In *Proc. ICASSP'96*, Atlanta, GA, May 1996.

[3] M. K. Tsatsanis and G. B. Giannakis, "Multirate filter banks for code-division multiple-access systems," In *Proc. ICASSP-95*, Detroit, MI, May 1995.

[4] M. K. Tsatsanis, "Inverse filtering criteria for CDMA systems," *IEEE Trans. on Signal Processing*, **45**(1):102–112, January 1997.

[5] Z. Zvonar and D. Brady, "Suboptimum multiuser detector for frequency-selective Rayleigh fading synchronous CDMA channels," *IEEE Trans. on Communications*, **43**:154–157, February/March/April 1995.

[6] D. H. Johnson and D. E. Dudgeon, *Array Signal Processing: Concepts and Techniques*, second edition, Englewood Cliffs, NJ: Prentice Hall, 1993.

[7] S. Haykin, *Adaptive Filter Theory*, Englewood Cliffs, NJ: Prentice Hall, 1991.

[8] R. O. Schmidt, "Multiple emitter location and signal parameter estimation," In *Proc. RADC Spectral Estimation Workshop*, pages 243–258, Griffiss AFB, NY, 1979.

[9] B. Ottersten, R. Roy, and T. Kailath, "Signal waveform estimation in sensor array processing," In *Proc. 23rd Asilomar Conference on Signals, Systems, and Computers*, Volume 2, pages 787–791, Pacific Grove, California, November 1989.

[10] G. Golub and C. Van Loan, *Matrix Computations*, second edition, Baltimore, MD: Johns Hopkins University Press, 1984.

[11] S. L. Miller, "Training analysis of adaptive interference suppression for direct-sequence CDMA systems," *IEEE Trans. on Communications*, **44**(4):488–495, April 1996.

[12] G. Xu and T. Kailath, "Fast Estimation of Principal Eigenspace Using the Lanczos Algorithm," *SIAM Journal on Matrix Analysis and Applications*, **15**(3):974–994, July 1994.

Chapter 3

CDMA With Short Codes: Indirect Approaches

3.1 Introduction

The receivers described in Chapter 2 are "direct" in the sense that they are constructed directly from the data observations. Such approaches are adequate but often suboptimum for a perfectly parameterized problem. For example, in constructing the DRAKE receiver, only the SOI is completely parameterized while no assumption is made on the functional form of the MAI. As a result, the number of coefficients that need to be determined from the data is $L_c(L + L_c)$ for each user, which is quite large considering that the actual number of unknowns per user is only L_c, the number of channel coefficients.

In contrast with the partially parametric approach discussed in Chapter 2, a full parametric approach in CDMA signal reception first estimates all unknowns and then constructs multiuser detectors based on the reconstructed effective signature waveforms. When used properly, the two-step (indirect) approach can make better use of the available data and yield higher detection performance, provided that the system model is accurate. The trade-off, in most cases, is the higher complexity incurred in multiuser parameter estimation. Since for the CDMA system described in (2.3) the model mismatch mainly comes from multipath with delays beyond L_c chips, and since signal energy attenuates rapidly with the delay (distance), the model error can be controlled to be arbitrarily small by increasing the maximum channel length L_c. In most cases, the model form is a reasonable description of reality. For this reason, indirect approaches are very suitable for uplink CDMA applications where performance outweighs the complexity.

In this chapter, we are concerned with the problem of multiuser parameter estimation in CDMA communications. We will first examine the feasibility of closed-form multiuser channel coefficients estimation without the use of training sequence. To put this into perspective, the total number of parameters that need to be determined in a P-user CDMA is $P \times L_c$. This number can be fairly large in a loaded system. We shall derive an estimation scheme that provides estimates of the multipath channels with fixed complexity by exploiting the structure information of the data output. In particular, we show that the subspace of the data matrix contains sufficient information for unique determination of the channel coefficients. Based on this observation, a MUSIC-like [1, 2] algorithm is presented.

In some scenarios, other system imperfections such as carrier drifts and Doppler shifts may also cause distortion to the signature waveforms. When such effects are not negligible, these parameters must be estimated jointly with the unknown channels. To add to the complexity, the distortion caused by carrier offsets is nonlinear. Regular estimation methods such as the maximum likelihood estimation are prohibitively expensive in such applications. In the second half of this chapter, the subspace channel estimation algorithm is extended to handle channel AND carrier offsets. An analytic algorithm for joint channel and carrier offset estimates is derived. The algorithm first converts the multiuser estimation problem into parallel single-user problems and then analytically solves the resulting nonlinear multivariate optimization problems using a polynomial matrix projection property.

Our discussions in this chapter will focus on quasi-synchronous CDMA systems as in Chapter 2 – the asynchronous problem can be handled by viewing it as large synchronous problems. After reviewing the subspace concept, we will introduce the channel coefficient estimation algorithm and discuss its implementation and other related issues in Section 3.2. Applications of this approach for handling overloaded systems are presented in Section 3.2.3. Section 3.3 extends the algorithm to joint channel and carrier offset estimation. Throughout this chapter, examples and computer simulations will be provided to illustrate the performance of these approaches.

3.2 Channel Parameter Estimation

Recall from Chapter 2 the received chip-rate CDMA signals within a symbol period:

$$\mathbf{y}(k) = \sum_{i=1}^{P} \bar{\mathbf{w}}_i s_i(k) + \mathbf{n}(k) = \sum_{i=1}^{P} \mathbf{C}_i \mathbf{h}_i s_i(k) + \mathbf{n}(k)$$

The unknowns in $\mathbf{y}(k)$ are the channel, $\{\mathbf{h}_i\}_{i=1}^{P}$, and the transmitted symbols, $\{s_i(k)\}_{i=1}^{P}$. To construct a multiuser receiver based on the effective signature waveforms $\{\bar{\mathbf{w}}_i\}$, it is necessary to estimate the $P \times L_c$ channel coefficients from a given number of observations, $\{\mathbf{y}(k)\}_{k=1}^{N}$. Such renders a well-defined multiuser parameter estimation problem. Theoretically the problem can be tackled using the maximum likelihood estimator (MLE) [3, 4] – the parameter set can be determined by maximizing a likelihood function in high-dimensional space. This, however, generally involves multidimensional iterative searching, which has some well-known implementational difficulties. In the following, we seek a more practical, noniterative approach with lower complexity and maybe suboptimal performance.

Before we derive the algorithm, it is worth pointing out that even when $\{\bar{\mathbf{w}}_i\}$ is unstructured (i.e., no prior information about the effective signature waveform), the multiuser identification problem of form:

$$\mathbf{y}(k) = \sum_{i=1}^{P} \bar{\mathbf{w}}_i s_i(k) = \bar{\mathbf{W}} \mathbf{s}(k)$$

may still be solvable using some blind algorithms that exploit structure information of the signal. These include the constant modulus approach (CM) [5] and other property restoral algorithms [6, 7]. The term *property restoral* refers to the fact that these algorithms force the signal estimates to have certain structural properties the actual signals are known to possess (e.g., finite-alphabet). However, such methods are often iterative and their effectiveness in dealing with a large number of users is still to be tested.

In Chapter 2, we showed how the structure of $\bar{\mathbf{w}}_i$ can be exploited for direct construction of the receivers. Here we will again take advantage of this information for the identification of the channel coefficients. Instead of estimating all the unknowns simultaneously, we will reduce the multiuser problem into a set of more tractable single-user problems using a subspace decomposition. The subspace-based algorithm utilized here is reminiscent of the well-known MUSIC [1] algorithm for direction-of-arrival (DOA) estimation in antenna array applications.

3.2.1 Subspace Concept

For the sake of presentation simplicity, we temporarily ignore the noise term in this section. The accommodation of noisy data will be addressed in algorithm implementation.

Collecting N observation vectors into a matrix, a subspace decomposition can be performed on $\mathbf{Y}$ using a singular value decomposition (SVD) [8]:

$$\begin{aligned}\mathbf{Y} &= [\mathbf{y}(1) \cdots \mathbf{y}(N)] = \bar{\mathbf{W}}[\mathbf{s}(1) \cdots \mathbf{s}(N)] = \bar{\mathbf{W}}_{(L+L_c) \times P} \mathbf{S}_{P \times N} \\ &= \begin{pmatrix} \mathbf{U}_s & \mathbf{U}_o \end{pmatrix} \begin{pmatrix} \mathbf{\Sigma}_s & 0 \\ 0 & 0 \end{pmatrix} \begin{pmatrix} \mathbf{V}_s^H \\ \mathbf{V}_o^H \end{pmatrix}\end{aligned} \quad (3.1)$$

where the vectors in $\mathbf{U}_s$, associated with the P non-zero singular values, span the "signal subspace" defined by the columns of $\bar{\mathbf{W}}$; while the vectors in $\mathbf{U}_o$, associated with the zero singular values, span the "orthogonal subspace" that is the orthogonal complement of the signal subspace:

$$\mathbf{U}_o \perp \bar{\mathbf{W}} \Rightarrow \mathbf{U}_o^H \bar{\mathbf{w}}_i = \mathbf{0}, \quad i = 1, \cdots, P \quad (3.2)$$

Here, we assume that (1) $P < L$, (2) $\{\bar{\mathbf{w}}_i\}$ in $\mathbf{W}$ are linearly independent, and (3) $\mathbf{S}$ has more than P columns and is of full row rank; all of which are reasonable conditions considering the randomness of the symbol sequences and the multipath channels. The dimension of $\mathbf{U}_s$ and $\mathbf{U}_o$ are $(L_c + L) \times P$ and $(L_c + L) \times (L_c + L - P)$, respectively.

3.2.2 Closed-Form Estimation

Once the connection between the subspaces and the effective signature waveform is made, the structure of $\{\bar{\mathbf{w}}_i\}$ can be used to link the orthogonal subspace with the channel coefficient vectors.

Substituting $\bar{\mathbf{w}}_i = \mathbf{C}_i \mathbf{h}_i$ into (3.2) yields:

$$\mathbf{U}_o^H \mathbf{C}_i \mathbf{h}_i = \mathbf{0}, \quad i = 1, \cdots, P \qquad (3.3)$$

The above equation set has $(L_c + L - P)$ equations and L_c unknowns. Therefore, if $P \leq L$, (3.3) is generally overdetermined and has a unique nontrivial solution: $\mathbf{h}_i$ subject to $\|\mathbf{h}_i\| = 1$. In other words, the channel coefficients for the ith user can be determined within a scalar ambiguity.

In an interference-limited CDMA system, the number of active users P is usually smaller than the spreading gain. Equation (3.3) provides us with a vital way to identify the channel coefficient directly from the data matrix. An outline of the algorithm is given below.

1. Construct the data matrix $\mathbf{Y}$ as in (3.1).

2. Apply an SVD to $\mathbf{Y}$, or equivalently, an eigenvalue decomposition to $\mathbf{Y}\mathbf{Y}^H$ to obtain the orthogonal subspace $\mathbf{U}_o$.

3. For each user, estimate the channel vector $\mathbf{h}_i$ by solving the linear equation set in (3.3).

4. For signal recovery, reconstruct the effective signature waveform vectors $\{\bar{\mathbf{w}}_i\}$ and calculate a CDMA multiuser detector.

The algorithm here exploits the important fact that each effective signature waveform vector is a linear function of a preassigned unique code. This fact, coupled with the redundancy in the overall system, allows perfectly determination of $\bar{\mathbf{w}}_i$ on the noise-free data vectors. The major computational cost of this algorithm comes from the subspace decomposition of the data matrix. The recent development of fast subspace decomposition (FSD) techniques provide computationally efficient, easily parallelizable methods for data decomposition [9].

It is worth pointing out that although the development is based on a noise-free model, it is not hard to modify the algorithm to deal with noise.

In the presence of additive noise, the data matrix becomes:

$$\mathbf{Y} = \mathbf{WS} + \mathbf{E}$$

We can still apply the SVD to the noise-corrupted data matrix and obtain a subspace decomposition similar to (3.1):

$$\mathbf{Y} = \begin{pmatrix} \tilde{\mathbf{U}}_s & \tilde{\mathbf{U}}_o \end{pmatrix} \begin{pmatrix} \tilde{\mathbf{\Sigma}}_s & 0 \\ 0 & \tilde{\mathbf{\Sigma}}_o \end{pmatrix} \begin{pmatrix} \tilde{\mathbf{V}}_s^H \\ \tilde{\mathbf{V}}_o^H \end{pmatrix} \quad (3.4)$$

Clearly, only an estimate of the orthogonal subspace is available. Thus, steps 2 and 3 in the algorithm above should be replaced with the following procedure:

2. Apply an SVD to the data matrix $\mathbf{Y}$ to obtain the orthogonal subspace as the singular vectors associated with the $L_c + L - P$ least-significant singular values.

3. For each user, estimate the channel coefficient vector $\hat{\mathbf{h}}_i$ as the least-squares solution of:

$$\tilde{\mathbf{U}}_o^H \mathbf{C}_i \mathbf{h}_i = \mathbf{0}, \quad i = 1, \cdots, P \quad (3.5)$$

After the above procedure, a zero-forcing decorrelating detector [10] can be constructed based on the reconstructed effective signature waveform $\{\bar{\mathbf{w}}_i\}$. To implement the minimum mean-square error (MMSE) multiuser detector [11] on the other hand, both the signal power σ_s^2 and the noise power σ_n^2 need to be further estimated. Assuming that both noise and symbols in $\mathbf{S}$ are zero-mean $i.i.d.$, these parameters can be achieved by noticing that:

$$\lim_{N \to \infty} \frac{1}{N} \tilde{\mathbf{\Sigma}}_o^2 = \begin{pmatrix} \sigma_n^2 & & 0 \\ & \ddots & \\ 0 & & \sigma_n^2 \end{pmatrix}$$

$$\lim_{N \to \infty} \tilde{\mathbf{R}}_\mathbf{Y} = \bar{\mathbf{W}} \begin{pmatrix} \sigma_s^2 & & 0 \\ & \ddots & \\ 0 & & \sigma_s^2 \end{pmatrix} \bar{\mathbf{W}}^H + \sigma_n^2 \mathbf{I}$$

where $\tilde{\mathbf{R}}_{\mathbf{Y}} = \frac{1}{N}\mathbf{Y}\mathbf{Y}^H$ is the sample data covariance matrix. The noise power can be estimated from the least-significant eigenvalue of $\tilde{\mathbf{R}}_{\mathbf{Y}}$; and with the effective signature waveform vector estimates $\hat{\tilde{\mathbf{W}}}$, the signal power and gains can be determined by:

$$\begin{pmatrix} \hat{\sigma}_s^2 & & 0 \\ & \ddots & \\ 0 & & \hat{\sigma}_s^2 \end{pmatrix} = \hat{\tilde{\mathbf{W}}}^\dagger \tilde{\mathbf{R}}_{\mathbf{Y}} \hat{\tilde{\mathbf{W}}}^{\dagger^H} - \hat{\sigma}_n^2 \mathbf{I}$$

where † denotes the left pseudo-inverse.

3.2.3 Overloaded Systems

The subspace method is capable of estimating up to L channel vectors in a CDMA system. In some applications, additional diversities (e.g., antenna arrays) are available and the system capacity can be further increased (see, for example, [12, 13, 14]). When the number of active users in the system exceeds L, direct application of the subspace algorithm becomes problematic. In particular, the algorithm breaks down as the number of users increases and the dimension of the orthogonal subspace $\mathbf{U}_o$ reduces to below L_c, the number of unknowns to be determined. To restore the identifiability, more equations are needed. In the following, we modify the subspace algorithm to accommodate more users by incorporating the spatial diversity.

Assume M receivers at the base station and denote superscript $()^m$ as the receiver index. The received signals at these antennas can be expressed as:

$$\mathbf{y}^m = \sum_{i=1}^{P} \bar{\mathbf{w}}_i^m s_i(k) + \mathbf{n}^m(k) = \sum_{i=1}^{P} \mathbf{C}_i \mathbf{h}_i^m s_i(k) + \mathbf{n}^m(k), \quad m = 1 \cdots M$$

We may stack the data vectors (matrices), the effective signature waveform vectors, and the channel vectors from all receivers in the following fashion:

$$\mathbf{y}(k) = \begin{bmatrix} \mathbf{y}^1(k) \\ \vdots \\ \mathbf{y}^M(k) \end{bmatrix} \quad \mathbf{Y} = \begin{bmatrix} \mathbf{Y}^1 \\ \vdots \\ \mathbf{Y}^M \end{bmatrix} \quad (3.6)$$

$$\bar{\mathbf{w}}_i = \begin{bmatrix} \bar{\mathbf{w}}_i^1 \\ \vdots \\ \bar{\mathbf{w}}_i^M \end{bmatrix} \quad \mathbf{h}_i = \begin{bmatrix} \mathbf{h}_i^1 \\ \vdots \\ \mathbf{h}_i^M \end{bmatrix} \quad (3.7)$$

Clearly, the input-output expressions:

$$\mathbf{y}(n) = \sum_{i=1}^{P} \bar{\mathbf{w}}_i s(n) \quad \text{and} \quad \mathbf{Y} = \bar{\mathbf{W}} \mathbf{S}$$

still hold, as does the subspace space relation between $\mathbf{Y}$ and $\bar{\mathbf{W}}$. However, the number of orthogonal vectors in $\mathbf{U}_o$ has been substantially increased to $M(L_c+L)-P$. At the same time, the signature vector $\bar{\mathbf{w}}_i$ is defined by the ML-channel vector $\mathbf{h}_i$ through a new kernel matrix:

$$\bar{\mathbf{w}}_i = diag \underbrace{(\mathbf{C}_i \cdots \mathbf{C}_i)}_{M \ blocks} \mathbf{h}_i$$

It is readily seen that $\{\mathbf{h}_i\}$ are given by the solution of the following linear equation sets:

$$\mathbf{U}_o^H \begin{bmatrix} \mathbf{C}_i & & 0 \\ & \ddots & \\ 0 & & \mathbf{C}_i \end{bmatrix} \mathbf{h}_i = \mathbf{0}; \quad i = 1, \cdots, P \quad (3.8)$$

Due to the additional diversity provided by the multiple receivers, the number of equations vs. the number of unknowns is now $M(L_c+L)-P : ML_c$, which means that in theory, $P \leq ML$ channels can be uniquely determined. Evidently, the more receivers the base station employs, the more channel vectors can be estimated, and the more users the CDMA system can handle. This is supported by the theoretical analysis on the relationship between the number of receivers and the system capacity [13].

The corresponding changes in the implementation procedure are straightforward.

3.2.4 Performance Analysis

While the subspace algorithm provides the exact channel estimates in the absence of noise, its estimates are inevitably perturbed when the data is noisy. Performance analysis is an important step before an estimation algorithm can be used in practice. In this particular application, the performance of CDMA signal recovery depends heavily on the accuracy of the reconstructed effective signature waveforms.

There is a large collection of papers on performance analysis of subspace-based parameter estimation algorithms. Here, we derive the bias and MSE expression of the signature waveform estimates using the first order perturbation theory introduced in [15]:

Lemma 1 *Let:*

$$\mathbf{X} = \begin{pmatrix} \mathbf{U}_s & \mathbf{U}_o \end{pmatrix} \begin{pmatrix} \mathbf{\Sigma}_s & 0 \\ 0 & 0 \end{pmatrix} \begin{pmatrix} \mathbf{V}_s^H \\ \mathbf{V}_o^H \end{pmatrix} \quad (3.9)$$

be the SVD of $\mathbf{X}$, *and* $\mathbf{E}$ *be an additive perturbation to* $\mathbf{X}$. *The first order approximation of the perturbation to* $\mathbf{U}_o$ *is given:*

$$\Delta \mathbf{U}_o = \mathbf{U}_s \mathbf{\Sigma}_s^{-1} \mathbf{V}_s^H \mathbf{E}^H \mathbf{U}_o = -\mathbf{X}^\dagger \mathbf{E}^H \mathbf{U}_o \quad (3.10)$$

For the problem under consideration, the noise-corrupted data matrix can be written as:

$$\mathbf{Y} = \mathbf{X} + \mathbf{E} = \mathbf{WS} + \mathbf{E}$$

The perturbed orthogonal subspace $\tilde{\mathbf{U}}_o$ in (3.4) can be expressed as the noise-free orthogonal subspace plus a perturbation term:

$$\tilde{\mathbf{U}}_o = \mathbf{U}_o + \Delta \mathbf{U}_o$$

By Lemma 1, the perturbation of the orthogonal subspace $\mathbf{U}_o$ can be expressed as $\Delta \mathbf{U}_o = -\mathbf{X}^\dagger \mathbf{E}^H \mathbf{U}_o$. Denoting $\mathbf{T}_i = \mathbf{C}_i^H \mathbf{U}_o$, it immediately follows that:

$$\Delta \mathbf{T}_i = \mathbf{C}_i^H \Delta \mathbf{U}_o = -\mathbf{C}_i^H \mathbf{X}^\dagger \mathbf{E}^H \mathbf{U}_o \quad (3.11)$$

Since in noise-free case, the $\mathbf{h}_i$ estimate is obtained as the unitary null vector of $\mathbf{T}_i$, we can apply Lemma 1 again and get the perturbation of the channel estimate:

$$\Delta \mathbf{h}_i = -\mathbf{T}_i^\dagger \Delta \mathbf{T}_i^H \mathbf{h}_i \quad (3.12)$$

Substituting (3.11) into the above equation yields:

$$\Delta \mathbf{h}_i = \mathbf{T}_i^\dagger \mathbf{U}_o^H \mathbf{E} \mathbf{X}^{\dagger^H} \underbrace{\mathbf{C}_i \mathbf{h}_i}_{\bar{\mathbf{w}}_i} \qquad (3.13)$$

Consequently, the perturbation of the signature waveform estimate is given by:

$$\Delta \bar{\mathbf{w}}_i = \mathbf{C}_i \Delta \mathbf{h}_i = \mathbf{C}_i \mathbf{T}_i^\dagger \mathbf{U}_o^H \mathbf{E} \mathbf{X}^{\dagger^H} \bar{\mathbf{w}}_i \qquad (3.14)$$

The noise elements in $\mathbf{E}$ are assumed to be zero-mean $i.i.d.$ Since in first-order approximation $\Delta \bar{\mathbf{w}}_i$ is linear with respect to $\mathbf{E}$, it is clear that the bias of the estimated signature waveform is zero.

To obtain the MSE of the effective signature waveform estimate, denote $\Delta \bar{\mathbf{w}}_i(k)$ the kth element of $\Delta \bar{\mathbf{w}}_i$ and $\mathbf{e}_j$ a column vector with the jth element 1 and all the rest 0, we have $\Delta \bar{\mathbf{w}}_i(k) = \mathbf{e}_k^H \Delta \bar{\mathbf{w}}_i$. Therefore,

$$E\left(\Delta \bar{\mathbf{w}}_i^H(k) \Delta \bar{\mathbf{w}}_i(k)\right) = E\left(\|\mathbf{e}_k^H \mathbf{C}_i \mathbf{T}_i^\dagger \mathbf{U}_o^H \mathbf{E} \mathbf{X}^{\dagger^H} \bar{\mathbf{w}}_i\|^2\right) \qquad (3.15)$$

$$= \sigma_n^2 \|\mathbf{e}_k^H \mathbf{C}_i \mathbf{T}_i^\dagger\|^2 \|\bar{\mathbf{w}}_i^H \mathbf{X}^\dagger\|^2 \qquad (3.16)$$

where the second equation is due to (38) in [15] and the fact that $\mathbf{U}_o^H \mathbf{U}_o = \mathbf{I}$. Also in [15] it is shown that $\|\bar{\mathbf{w}}_i^H \mathbf{X}^\dagger\|^2 = \frac{1}{N\sigma_s^2}$. Thus,

$$E\left(\Delta \bar{\mathbf{w}}_i^H \Delta \bar{\mathbf{w}}_i\right) = \frac{\sigma_n^2}{N\sigma_i^2} \sum_k \|\mathbf{e}_k^H \mathbf{C}_i \mathbf{T}_i^\dagger\|^2 \qquad (3.17)$$

Since $\sum_k \|\mathbf{e}_k^H \mathbf{C}_i \mathbf{T}_i^\dagger\|^2 = \|\mathbf{C}_i \mathbf{T}_i^\dagger\|^2$, we obtain the final expression of the MSE for the signature waveform estimates.

$$E\left(\|\Delta \bar{\mathbf{w}}_i\|^2\right) = \frac{\sigma_n^2 \|\mathbf{C}_i \mathbf{T}_i^\dagger\|^2}{\sigma_s^2 |\gamma_i|^2 N} \qquad (3.18)$$

Some examples are presented here to illustrate the performance of the algorithm quantitatively. In all of the following examples, the multipath delay and the number of multipath components are uniformly distributed within $[0 \quad 3T]$ and $[1, 10]$, respectively. Power control within 3 dB is assumed.

CDMA With Short Codes: Indirect Approaches

Example 1: The performance improvement due to incorporating the channel estimates in detection is examined in the first case. We simulate a CDMA system with $L = 32$, $P = 25$, and SNR= 5 dB. The users' codes are randomly generated. $N = 80$ data vectors are used for channel estimation, although in principle only $N \geq P$ sample vectors are required to guarantee that $\mathbf{S}$ is of full row rank. L_c is preselected to be 4. After $\{\mathbf{h}_i\}$ are determined using the subspace approach, the effective signature vectors $\{\bar{\mathbf{w}}_i\}$ are reconstructed. We then compute the simple zero-forcing decorrelating receiver to recover the original signal for each user. Figure 3.1 illustrates the channel responses used in the simulations and processing results for one of the users. Comparing the signal constellations using the zero-forcing equalizer and conventional matched filter, the performance gain due to multiuser detection based on the effective signature waveform estimates is significant.

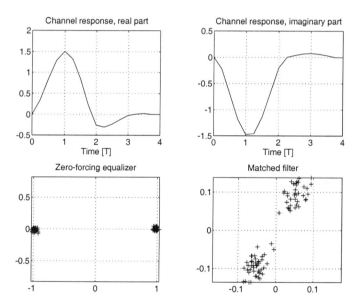

Figure 3.1 Channel and signal constellation of user 1.

Figure 3.2 shows the same results for another user who experiences stronger and longer delay multipath. Again, the subspace algorithm effectively determines the signature waveform for the recovery of the message signals. The result indicates that the subspace algorithm is robust against channel conditions so long as the channel length is within L_c chips.

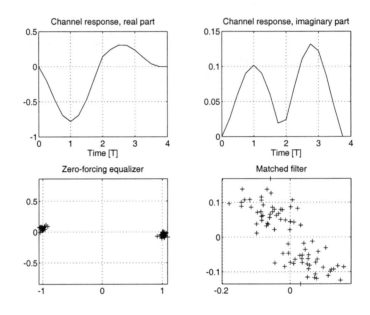

Figure 3.2 Channel and signal constellation of user 2.

Example 2: The next example is given primarily to demonstrate how well the theoretical expressions predict the performance of the effective signature waveform estimates under different SNRs. A total of 500 trials are conducted with SNR varying from 0 to 30. The sample root MSE of the effective signature waveform estimates are then calculated and compared to that predicted by the corresponding theoretical expressions. The results of the simulations for the first user are plotted in Figure 3.3. The symbol o represents the sample RMSE of the signature waveform estimates. The solid line represents the error predicted by the analysis. Note the excellent agreement between predicted and simulated values, even for the left-most case where the SNR is very low.

Example 3: As an application of our approach to an overloaded CDMA, we consider a two-receiver system with 50 users. The rest of the setup remains the same as that of previous examples. The SINR distribution of all users' symbol sequences after channel estimation and zero-forcing reception is plotted in Figure 3.4. To some extent, this verifies the capability of the subspace approach and the performance enhancement promised by the multireceiver, multiuser detection techniques.

3.3 Joint Channel and Carrier Estimation

While the channel effect is the primary distorting factor in wideband CDMA communications, in some applications carrier offsets caused by carrier drafts or Doppler shifts may also contribute to the distortion of users' signature waveforms. In a multiple-access environment, the effect due to different carrier offsets from different CDMA users cannot be compensated individually. Parameter estimation followed by multiuser detection is the only cure to this problem.

In this section, we will extend the subspace algorithm to handle joint channel and carrier offset estimation. Our goal is to derive an approach that provides analytic solutions using available numerical computing tools such as the SVD and polynomial root algorithms.

Let us first establish the data model for the CDMA signal with multipath channel and carrier offset effects. Introducing carrier offsets to the data model, the received CDMA signal with multipath channel and residual carrier offsets is given by:

$$y(k) = \sum_{i=1}^{P} y_i(k) + n(k) = \sum_{i=1}^{P} \sum_{n=-\infty}^{\infty} x_i(n) e^{j\phi_i k} h_i(n-k) + n(k) \quad (3.19)$$

ϕ_i here represents the carrier offset associated with the ith user. When $\{\phi_i\}_{i=1}^{P}$ are negligible, the above expression reduces to the channel-only expression we are familiar with. Otherwise, the multiplicative effect of the residual carriers has to be dealt with explicitly.

Through straightforward manipulations, we can rearrange $y(k)$

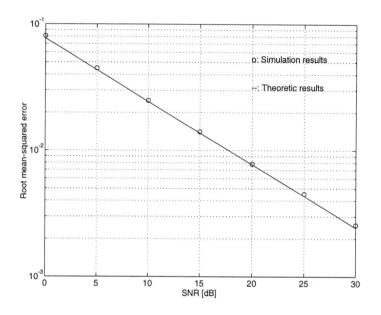

Figure 3.3 Root mean-square error vs. SNR for user 1.

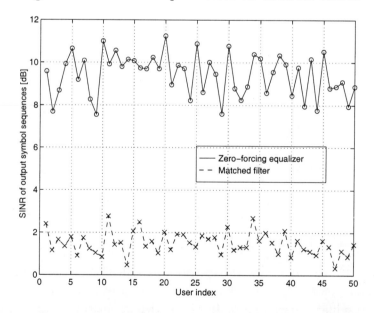

Figure 3.4 SINR distribution of output symbol sequences.

CDMA With Short Codes: Indirect Approaches

within each symbol period in vector form as:

$$\mathbf{y}(k) = \sum_{i=1}^{P} s_i(k) e^{jkM\phi_i} \begin{bmatrix} \bar{w}_i(1) \\ \bar{w}_i(2) \\ \vdots \\ \bar{w}_i(L+L_c) \end{bmatrix} + \mathbf{n}(k)$$

$$\stackrel{\text{def}}{=} \sum_{i=1}^{P} s_i(k) e^{jkM\phi_i} \bar{\mathbf{w}}_i + \mathbf{n}(k) \qquad (3.20)$$

The effective signature waveform vector $\bar{\mathbf{w}}_i$ is now a function of both the channel response and the carrier offset:

$$\bar{\mathbf{w}}_i = \underbrace{\begin{bmatrix} 1 & 0 & \cdots & 0 \\ 0 & e^{j\phi_i} & \cdots & 0 \\ \vdots & & \ddots & \vdots \\ 0 & 0 & \cdots & e^{j\phi_i(L+L_c)} \end{bmatrix}}_{\mathbf{Z}_i} \mathbf{C}_i \mathbf{h}_i = \mathbf{Z}_i \mathbf{C}_i \mathbf{h}_i \qquad (3.21)$$

The general problem addressed here is the estimation of $\{\bar{\mathbf{w}}_i\}$ or $\{\phi_i\}$ and $\{\mathbf{h}_i\}$ from $\{\mathbf{y}(k)\}_{k=1}^{N}$ without the knowledge of the information symbol $\{s_i(k)\}$.

Again, the algebraic structure revealed in $\bar{\mathbf{w}}_i = \mathbf{Z}_i \mathbf{C}_i \mathbf{h}_i$ plays an important role in solving the joint estimation problem. Once $\{\bar{\mathbf{w}}_i\}_{i=1}^{P}$ are estimated, multiuser detection can easily be performed using decorrelating or MMSE receivers [16–20].

3.3.1 Problem Reformulation

The present problem is clearly more involved than the channel-only case. Fortunately, the subspace approach that converts the multiuser parameter estimation problem into parallel single-user ones still apply. Our strategy here is as follows. First, the subspace approach will be used to reduce the multiuser problem into P single-user estimation problems through subspace decomposition; after that, each single-user problem will be simplified to a tractable 1-D minimization problem. As will become clear, the key to joint estimation is a polynomial matrix operation that enables us to decouple carrier and channel in the cost function.

Again, we will derive the algorithm using noise-free data. From (3.1) and (3.20),

$$\begin{aligned}
\mathbf{Y} &= [\mathbf{y}(1) \cdots \mathbf{y}(N)] \\
&= [\bar{\mathbf{w}}_1 \cdots \bar{\mathbf{w}}_P] \begin{bmatrix} s_1(1) & s_1(2)e^{jM\phi_1} & \cdots & s_1(N)e^{j(N-1)M\phi_1} \\ \vdots & & \ddots & \vdots \\ s_P(1) & s_P(2)e^{jM\phi_P} & \cdots & s_P(N)e^{j(N-1)M\phi_P} \end{bmatrix} \\
&\stackrel{\text{def}}{=} \bar{\mathbf{W}}\mathbf{S}
\end{aligned} \quad (3.22)$$

Applying an SVD to $\mathbf{Y}$ yields:

$$\mathbf{Y} = \bar{\mathbf{W}}\mathbf{S} = (\mathbf{U}_s \ \mathbf{U}_o) \begin{pmatrix} \Sigma_s & 0 \\ 0 & 0 \end{pmatrix} \begin{pmatrix} \mathbf{V}_s^H \\ \mathbf{V}_o^H \end{pmatrix} \quad (3.23)$$

Because of the subspace relation, the following homogeneous equations hold:

$$\mathbf{U}_o \perp \bar{\mathbf{W}} \Rightarrow \mathbf{U}_o^H \bar{\mathbf{w}}_i = \mathbf{U}_o^H \mathbf{Z}_i \mathbf{C}_i \mathbf{h}_i = 0, \quad i = 1, \cdots, P \quad (3.24)$$

Unlike the channel-only case, the parameters in $\{\bar{\mathbf{w}}_i\}$ can no longer be estimated directly because $\|\mathbf{U}_o^H \mathbf{Z}_i \mathbf{C}_i \mathbf{h}_i\|^2$ is not in quadratic form with respect to ϕ_i. On the other hand, (3.24) does reduce the multiuser parameter estimation to P single-user problems.

Upon defining $J(\phi_i, \mathbf{h}_i) = \|\mathbf{U}_o^H \mathbf{Z}_i \mathbf{C}_i \mathbf{h}_i\|^2$, one can estimate the carrier and channel of each user as:

$$\hat{\phi}_i, \hat{\mathbf{h}}_i = \underset{\phi_i, \mathbf{h}_i}{\operatorname{argmin}} \|J(\phi_i, \mathbf{h}_i)\|^2 \quad i = 1, \cdots, P \quad (3.25)$$

Solving the above type of multivariant minimization problem usually involves high-dimensional iterative searching. This is challenging because:

1. The estimation of ϕ_i and $\mathbf{h}_i$ cannot be easily decoupled.

2. $\mathbf{Z}_i(\phi_i) = \operatorname{diag}[1, e^{j\phi_i}, \cdots, e^{\phi_i(M-1)}]$ is nonlinearly dependent on ϕ_i.

CDMA With Short Codes: Indirect Approaches

To obtain an analytic solution, we need to convert $J(\phi, \mathbf{h})$ into a more tractable form.

For clarity, we drop the subscript in (3.24) and consider the following:
$$\mathbf{U}_o^H \mathbf{Z} \mathbf{C} \mathbf{h} = \mathbf{0} \quad (3.26)$$

Defining $z = e^{j\phi}$, (3.26) can be partitioned as:

$$[\mathbf{u}_{o1} \ \mathbf{u}_{o2} \ \cdots \ \mathbf{u}_{oK}] \begin{bmatrix} 1 & 0 & \cdots & 0 \\ 0 & z & \cdots & 0 \\ \vdots & & \ddots & \vdots \\ 0 & 0 & \cdots & z^{K-1} \end{bmatrix} \begin{bmatrix} \mathbf{c}_1^T \\ \mathbf{c}_2^T \\ \vdots \\ \mathbf{c}_K^T \end{bmatrix} \mathbf{h} = \mathbf{0} \quad (3.27)$$

where $\mathbf{u}_{oi}$ is the ith column vector of $\mathbf{U}_o^H$, and $\mathbf{c}_j^T$ is the jth row vector of $\mathbf{C}$. Note that the product $\mathbf{U}_o^H \mathbf{Z} \mathbf{C}$ in the above equation can be further expressed as the following polynomial matrix with variable z,

$$[\mathbf{u}_{o1} \ \mathbf{u}_{o2} \ \cdots \ \mathbf{u}_{oK}] \begin{bmatrix} 1 & 0 & \cdots & 0 \\ 0 & z & \cdots & 0 \\ \vdots & & \ddots & \vdots \\ 0 & 0 & \cdots & z^{K-1} \end{bmatrix} \begin{bmatrix} \mathbf{c}_1^T \\ \mathbf{c}_2^T \\ \vdots \\ \mathbf{c}_K^T \end{bmatrix} = \underbrace{\sum_{i=1}^{K} \mathbf{u}_{oi} \mathbf{c}_i^T z^{i-1}}_{\mathbf{Q}(z)}$$

(3.28)

$\mathbf{Q}(z)$ is an $L \times L$ polynomial matrix of order $K - 1$. The parameter estimation problem in (3.25) is equivalent to the minimization of the following quantity:

$$\|\mathbf{Q}(z)\mathbf{h}\| = \|[\mathbf{q}_1(z), \cdots, \mathbf{q}_L(z)]\mathbf{h}\| \quad (3.29)$$

Clearly, $\mathbf{h}$ is a null vector of $\mathbf{Q}(z)$ when $z = e^{j\phi_0}$, where ϕ_0 is the true carrier offset. The nullity of $\mathbf{Q}(e^{j\phi_0})$ is thus at least 1. Barring degeneration cases, this suggests ϕ_0 can be determined through examining the rank condition of $\mathbf{Q}(z)$ along the unit circle. In other words, we may decouple the joint carrier offset and channel estimation by finding ϕ_o as the frequency at which the nullity of $\mathbf{Q}(z)$ is one. More specifically,

- The carrier offset is estimated as the frequency at which the nullity of $\mathbf{Q}(z)$ is one.

- After the carrier offset is determined, the channel coefficient is found as the least-square solution that minimizes the cost function in (3.29).

3.3.2 Decoupling and Closed-Form Solution

There exist no numerical tools that can be readily applied to evaluate the rank condition of a matrix polynomial. In the following, we present an algorithm that solves this problem to simple polynomial rooting or one-dimensional spectrum searching.

To evaluate the nullity of $\mathbf{Q}(z)$, let:

$$\mathbf{F}(z) = [\mathbf{q}_2(z), \cdots, \mathbf{q}_L(z)] \tag{3.30}$$

and note from linear algebra that if a vector $\mathbf{x}$ lies in the column span of a matrix $\mathbf{A}$, $\mathbf{P}_A^\perp \mathbf{x} = 0$. Here $\mathbf{P}_A^\perp$ is the orthogonal projection matrix of $\mathbf{A}$. Therefore, if we denote $\mathbf{P}_F^\perp(z)$ as the orthogonal projection polynomial matrix of $\mathbf{F}(z)$, $\mathbf{P}_F^\perp(z)\mathbf{q}_1(z) = 0$ when the nullity of $\mathbf{Q}(z)$ is nonzero. In other words, ϕ_0 can be solved as[1]:

$$\phi_0 = \underset{\phi}{\operatorname{argmin}} \, \|\mathbf{P}_F^\perp(z)\mathbf{q}_1(z)\| \tag{3.31}$$

The issue now is whether one can construct the explicit form of the orthogonal projection of a polynomial matrix. To find $\mathbf{P}_F^\perp(z)$, we first need to construct a polynomial matrix $\mathbf{G}(z)$ satisfying:

$$\mathbf{G}(z)\mathbf{F}(z) = z^{-n_0}\mathbf{I} \tag{3.32}$$

where n_0 is an appropriate delay index. Then $\mathbf{F}(z)\mathbf{G}(z)$ is a legitimate projection matrix of $\mathbf{F}(z)$ with delay n_0. Such is indeed possible when $\mathbf{F}(z)$ is a full-rank tall matrix. In Appendix 3A, we show how an FIR $\mathbf{G}(z)$ can be constructed directly from $\mathbf{F}(z)$. Discussions on the existence of such an FIR inverse are also provided.

---

[1] Theoretically, when $\mathbf{Q}(z)$ is rank deficient, it is possible for $\mathbf{q}_1(z)$ to be independent with $\mathbf{F}(z)$ while $\mathbf{F}(z)$ itself is rank deficient. In practice, however, the projection of a randomly picked column (e.g., $\mathbf{q}_1(z)$) to the rest of the columns of $\mathbf{Q}(z)$ can almost surely reveal its rank condition.

CDMA With Short Codes: Indirect Approaches

Once $\mathbf{G}(z)$ is constructed, the orthogonal projection matrix for $\mathbf{F}(z)$, which is still an FIR polynomial matrix, is given by:

$$\mathbf{P}_F^{\perp}(z) = (z^{-n_0}\mathbf{I} - \mathbf{F}(z)\mathbf{G}(z)) \quad (3.33)$$

Substituting (3.33) into (3.31) and defining:

$$\mathbf{p}(z) \overset{\text{def}}{=} \mathbf{P}_F^{\perp}(z)\mathbf{q}_1(z) = (z^{-n_0}\mathbf{I} - \mathbf{F}(z)\mathbf{G}(z))\mathbf{q}_1(z), \quad (3.34)$$

$z_0 = e^{j\phi_0}$ is the solution of the following minimization problem:

$$z_0 = \underset{z}{\text{argmin}} \, \|\mathbf{p}(z)\|^2 \quad (3.35)$$

Let $a(z) = \|\mathbf{p}(z)\|^2 = \mathbf{p}^H(z)\mathbf{p}(z)$, ϕ_0 is readily determined by evaluating $a(z)$ along the unit circle. At the same time, it is noticed that $a(z)$ itself forms a polynomial with $z_0 = e^{j\phi_0}$ as one of its roots. This suggests an alternative way to identify ϕ_0 through polynomial rooting as in the Root-MUSIC algorithm [21]. In practice, one often picks the root inside the unit circle with the largest magnitude. Its angle is the estimate of ϕ_0.

Graphical illustrations of the cost function $a(z) = \|\mathbf{p}(z)\|^2$ in (3.34) using noise-free data are given in Figure 3.5 and Figure 3.6. Figure 3.5 shows the spectrum of $a(\phi)$, while Figure 3.6 plots the root distribution of $a(z)$ of a 10-user CDMA system with spreading gain $L = 32$. In Figure 3.5, the solid line denotes the true carrier offset. The "o" in Figure 3.6 represents the root corresponding to the ϕ estimate. As seen, the spectrum yields the exact carrier offset estimate in the absence of noise. The sharp spectrum also indicates good statistical properties of the algorithm. Also noticed in the spectrum are the many local minima. Fortunately, in practice, one only needs to search within a small vicinity to locate the carrier offset.

On the other hand, the polynomial rooting method may be too sensitive to be implemented due to the large number of zeros around the unit circle. Spectrum searching within the vicinity of the carrier offset is more practical for this application.

Once $z = e^{j\phi}$ is determined, the channel $\mathbf{h}$ can be estimated from (3.24) as:

$$\hat{\mathbf{h}} = \underset{\|\mathbf{h}=1\|}{\text{argmin}} \, \mathbf{h}^H \mathbf{C}^H \hat{\mathbf{Z}}^H \mathbf{U}_o \mathbf{U}_o^H \hat{\mathbf{Z}} \mathbf{C} \mathbf{h} \quad (3.36)$$

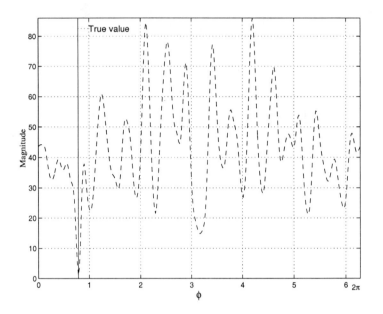

Figure 3.5 Null spectrum.

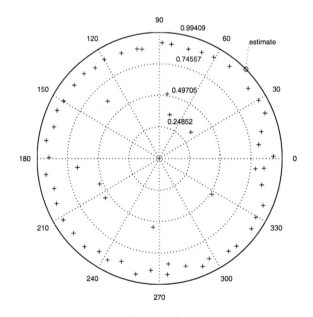

Figure 3.6 Root distribution.

That is, the channel estimate is given by the null vector of $\mathbf{U}_o^H \hat{\mathbf{Z}} \mathbf{C}$ corresponding to the least-significant singular value.

The joint channel and carrier offset estimation algorithm is summarized below:

1. Perform an SVD on the data matrix $\mathbf{Y}$ to obtain the orthogonal subspace $\mathbf{U}_o$.
2. For each user, form the matrix polynomial $\mathbf{F}_i(z)$ as in (3.30).
3. Calculate the FIR inverse of $\mathbf{F}_i(z)$ and then the orthogonal projection matrix $\mathbf{P}_{F_i}^\perp(z)$.
4. Identify the carrier offset $z_i = e^{j\phi_i}$ from cost function $a_i(z) = \|\mathbf{p}_i(z)\|^2$ through spectrum searching or polynomial rooting.
5. Construct $\hat{\mathbf{Z}}_i(z_i)$ and determine the channel vector $\mathbf{h}_i$ as the least-square solution of (3.36).

Once the channel response vectors and carrier offsets are estimated, we can form a decorrelating receiver to recover each user's information bearing symbols. Multiuser detectors can be constructed simultaneously based on the effective signature waveforms:

$$\mathbf{w}_i = \mathbf{Z}_i \mathbf{C}_i \mathbf{h}_i, \quad i = 1 \cdots P$$

3.3.3 Performance Analysis

Similar first-order perturbation analysis can be adopted to examine the performance of the joint estimation algorithm.

The algorithm relies on minimization of the cost function that reaches its minimum at the true ϕ_i and $\mathbf{h}_i$. Since the channel vector can only be identified within a scalar ambiguity, in blind identification we often impose a constraint such that the estimate is unique. In our analysis, we will assume the first coefficient of $\mathbf{h}$ is known. This is equivalent to a linear constraint:

$$\mathbf{e}_1^H \mathbf{h} = 1$$

where $\mathbf{e}_1$ is the first column of an identity matrix with proper dimension. It should be pointed out that the actual algorithm is based

70 *Signal Processing Applications in CDMA Communications*

on a quadratic constraint. However, within first order, the difference between these two constraints should be small.

Denote $\mathbf{h}_o$ as $\mathbf{h}$ without the first element and $\check{\mathbf{C}}$ as $\mathbf{C}$ without the first column. Further define:

$$\eta = [\phi, \bar{\mathbf{h}}_o^T, \tilde{\mathbf{h}}_o^T]^T$$

where $(\bar{\cdot})$ and $(\tilde{\cdot})$ denote the real time and the imaginary part of $(\cdot)$, respectively. When $J(\hat{\phi}, \hat{\mathbf{h}}, \hat{\mathbf{U}}_o)$ of the noisy data reaches its minimum, its first-order derivative with respect to ϕ and $\mathbf{h}_o$ must be 0. Approximating the first-order partial derivatives of J at $(\hat{\phi}, \hat{\mathbf{h}}, \hat{\mathbf{U}}_o)$ using first-order Taylor series expansion at point $(\phi, \mathbf{h}, \hat{\mathbf{U}}_o)$ yields:

$$0 = \frac{\partial J(\hat{\phi}, \hat{\mathbf{h}}, \hat{\mathbf{U}}_o)}{\partial \eta} = \frac{\partial J(\phi, \mathbf{h}, \hat{\mathbf{U}}_o)}{\partial \eta} + \frac{\partial^2 J(\phi, \mathbf{h}, \hat{\mathbf{U}}_o)}{\partial \eta^2} \Delta \eta \qquad (3.37)$$

The second order-partial derivative is given by:

$$\frac{\partial^2 J(\phi, \mathbf{h}, \hat{\mathbf{U}}_o)}{\partial \eta^2} = \begin{bmatrix} \frac{\partial^2 J}{\partial \phi^2} & \frac{\partial^2 J}{\partial \phi \partial \bar{\mathbf{h}}_o} & \frac{\partial^2 J}{\partial \phi \tilde{\mathbf{h}}_o} \\ \frac{\partial^2 J}{\partial \bar{\mathbf{h}}_o \partial \phi} & \frac{\partial^2 J}{\partial \bar{\mathbf{h}}_o^2} & \frac{\partial^2 J}{\partial \bar{\mathbf{h}}_o \partial \tilde{\mathbf{h}}_o} \\ \frac{\partial^2 J}{\partial \tilde{\mathbf{h}}_o \partial \phi} & \frac{\partial^2 J}{\partial \tilde{\mathbf{h}}_o \partial \bar{\mathbf{h}}_o} & \frac{\partial^2 J}{\partial \tilde{\mathbf{h}}_o^2} \end{bmatrix} \stackrel{\text{def}}{=} \mathbf{H}_J \qquad (3.38)$$

From (3.37), we can solve for the perturbation of the parameter estimates, $\Delta \phi$, $\Delta \bar{\mathbf{h}}$, and $\Delta \tilde{\mathbf{h}}$, as:

$$\Delta \eta = \begin{bmatrix} \Delta \phi \\ \Delta \bar{\mathbf{h}}_o \\ \Delta \tilde{\mathbf{h}}_o \end{bmatrix} = -\mathbf{H}_J^{-1} \begin{bmatrix} \frac{\partial J(\phi, \mathbf{h}, \hat{\mathbf{U}}_o)}{\partial \phi} \\ \frac{\partial J(\phi, \mathbf{h}, \hat{\mathbf{U}}_o)}{\partial \bar{\mathbf{h}}_o} \\ \frac{\partial J(\phi, \mathbf{h}, \hat{\mathbf{U}}_o)}{\partial \tilde{\mathbf{h}}_o} \end{bmatrix} \qquad (3.39)$$

Given the above expression, we next express $\Delta \eta$ as a linear function of the first-order perturbation of the orthogonal subspace $\Delta \mathbf{U}_o$. For notational simplicity, we denote $\frac{\partial A(\phi, \mathbf{h})}{\partial \phi}$ as $A'(\phi, \mathbf{h})$ and define the first-order partial derivative $\mathbf{D} = \frac{\partial J}{\partial \eta}$. We have, from (3.39):

$$\Delta \eta = -\mathbf{H}_J(\phi, \mathbf{h}, \hat{\mathbf{U}}_o)^{-1} \mathbf{D}(\phi, \mathbf{h}, \hat{\mathbf{U}}_o) \qquad (3.40)$$

CDMA With Short Codes: Indirect Approaches

The first- and second-order partial derivatives that appeared in (3.39) are calculated in Appendix 3B.

When the noise power is small relative to the signal strength, we can use first-order approximation and obtain:

$$\mathbf{D}(\phi, \mathbf{h}, \hat{\mathbf{U}}_o) \approx \mathbf{D}(\phi, \mathbf{h}, \mathbf{U}_o) + \Delta \mathbf{D}(\phi, \mathbf{h}, \mathbf{U}_o)$$

and

$$\mathbf{H}_J(\phi, \mathbf{h}, \hat{\mathbf{U}}_o) \approx \mathbf{H}_J(\phi, \mathbf{h}, \mathbf{U}_o) + \Delta \mathbf{H}_J(\phi, \mathbf{h}, \mathbf{U}_o)$$

Note $\mathbf{D}(\phi, \mathbf{h}, \mathbf{U}_o) = \mathbf{0}$, we can further approximate (3.40) as follows:

$$\Delta \eta \approx -(\mathbf{H}_J^{-1} - \mathbf{H}_J^{-1} \Delta \mathbf{H}_J \mathbf{H}_J^{-1}) \Delta \mathbf{D} \approx -\mathbf{H}_J^{-1} \Delta \mathbf{D} \quad (3.41)$$

Combining the above results with those in Appendix B, $\Delta \mathbf{D}$ can be expressed in terms of noise $\mathbf{N}$:

$$\begin{aligned}
-\Delta \mathbf{D}_\phi &= \mathbf{h}^H \mathbf{C}^H \mathbf{Z}_z'^H \mathbf{U}_o \mathbf{U}_o^H \mathbf{N} \mathbf{X}^\dagger \mathbf{Z} \mathbf{C} \mathbf{h} \\
&+ \mathbf{h}^H \mathbf{C}^H \mathbf{Z}^H \mathbf{X}^{\dagger H} \mathbf{N}^H \mathbf{U}_o \mathbf{U}_o^H \mathbf{Z}_z' \mathbf{C} \mathbf{h} \\
-\Delta \mathbf{D}_{\tilde{\mathbf{h}}_o} &= \check{\mathbf{C}}^H \mathbf{Z}^H \mathbf{U}_o \mathbf{U}_o^H \mathbf{N} \mathbf{X}^\dagger \mathbf{Z} \mathbf{C} \mathbf{h} \\
&+ \check{\mathbf{C}}^T \mathbf{Z}^T \mathbf{U}_o^* \mathbf{U}_o^T \mathbf{N}^* \mathbf{X}^{\dagger *} \mathbf{Z}^* \mathbf{C}^* \mathbf{h}^* \\
-\Delta \mathbf{D}_{\tilde{\mathbf{h}}_o} &= -j \check{\mathbf{C}}^H \mathbf{Z}^H \mathbf{U}_o \mathbf{U}_o^H \mathbf{N} \mathbf{X}^\dagger \mathbf{Z} \mathbf{C} \mathbf{h} \\
&+ j \check{\mathbf{C}}^T \mathbf{Z}^T \mathbf{U}_o^* \mathbf{U}_o^T \mathbf{N}^* \mathbf{X}^{\dagger *} \mathbf{Z}^* \mathbf{C}^* \mathbf{h}^*
\end{aligned} \quad (3.42)$$

When the noise is zero mean, it is easy to verify $E[\Delta \eta_i] = 0$. The joint estimator is thus unbiased. To obtain the MSEs of parameter estimates, we need to calculate $E[\Delta \eta_i^2]$. Assume the noise is independent complex white Gaussian with zero mean and variance of $2\sigma^2$. The following results hold:

$$\begin{aligned}
E[\mathbf{a}^H \mathbf{N} \mathbf{b} \mathbf{c}^H \mathbf{N} \mathbf{d}] &= 0 \\
E[\mathbf{a}^H \mathbf{N}^H \mathbf{b} \mathbf{c}^H \mathbf{N}^H \mathbf{d}] &= 0 \\
E[\mathbf{a}^H \mathbf{N} \mathbf{b} \mathbf{c}^H \mathbf{N}^H \mathbf{d}] &= 2 \mathbf{a}^H \mathbf{d} \mathbf{c}^H \mathbf{b} \sigma^2 \\
E[\mathbf{a}^H \mathbf{N} \mathbf{b} \mathbf{c}^H \mathbf{N}^* \mathbf{d}] &= 2 \operatorname{Re}[\mathbf{a}^H \mathbf{c}^* \mathbf{b}^T \mathbf{d} \sigma^2]
\end{aligned} \quad (3.43)$$

Define $\mathbf{T} = \mathbf{H}_J^{-1}$. Further rewrite $\Delta \mathbf{D}_{\eta_i}$ in (3.42) as $-\Delta \mathbf{D}_{\eta_i} = \alpha_i^H \mathbf{N} \beta_i + \gamma^H \mathbf{N}^H \delta_i$ or $-\Delta \mathbf{D}_{\eta_i} = \alpha_i^H \mathbf{N} \beta_i + \epsilon_i^H \mathbf{N}^* \zeta$ with α_i, β_i, γ_i, δ_i,

ϵ_i, and ζ_i accordingly defined. The final MSE expressions for each parameter estimate are

$$\begin{aligned}
E[\Delta \eta_i^2] = & \ 4\sigma^2 \operatorname{Re}\Big\{ \mathbf{T}(i,1)^2 \alpha_1^H \delta_1 \gamma_1^H \beta_1 \\
& + \sum_{j=2}^{(2PL-1)} \mathbf{T}(i,j)^2 \alpha_j^H \epsilon_j^* \beta_j^T \zeta_j \\
& + \sum_{j=2}^{(2PL-1)} (\mathbf{T}(i,1)\mathbf{T}(i,j)\alpha_1 \epsilon_j^* \beta_1^T \zeta_j \\
& + \mathbf{T}(i,j)\mathbf{T}(i,1)\alpha_j^H \delta_1 \gamma_1^H \beta_j) \\
& + \sum_{j=2}^{(2PL-2)} \sum_{k=j+1}^{(2PL-1)} 2 * \mathbf{T}(i,j)\mathbf{T}(i,k) \alpha_j^H \epsilon_k^* \beta_j^T \zeta_k \Big\} \\
& i = 1, \cdots, n
\end{aligned} \quad (3.44)$$

It is important to point out that though seemingly complicated, the above MSE expression only contains physical parameters of the system, thus allowing one to predict the performance of the algorithm directly by plugging in the system parameters, including the true carriers and channels.

3.4 Examples

In this section, we provide some computer simulation results to illustrate the efficacy of the joint estimation algorithm and verify the analytic MSE expression derived for the algorithm. In all of the following examples, multipath channels of length 3 are used. Carrier offsets are randomly selected from $[\frac{-\pi}{32}, \frac{\pi}{32}]$. The information sequence is QPSK.

Example 4: In this example, we verify the performance predicted by perturbation analysis in Section 3.3.3 under different SNRs with the following setup: $P=2$, $M=8$, $N=20$. With SNR varying from 10 dB to 30dB, 365 Monte Carlo trials are conducted for each SNR value. The MSE is calculated and compared to that given by theoretical prediction and the CR bound. We plot out the MSEs for ϕ_1 and $\mathbf{h}_1$ in Figures 3.7 and 3.8, respectively. The normalized MSE $\|\hat{\mathbf{h}}_1 - \mathbf{h}_1\|^2 / \|\mathbf{h}_1\|^2$

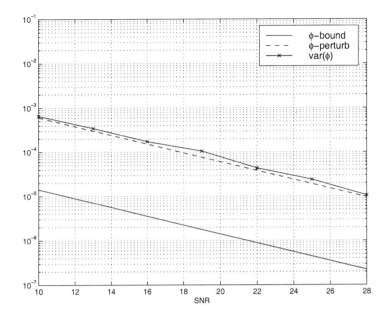

Figure 3.7 Performance of ϕ estimation.

is employed as the performance measure. As can be seen from the two figures, the MSEs of carrier and channel estimate obtained from simulation and perturbation analysis are very close, indicating the accuracy of the perturbation analysis for a large range of SNR values. The same results have been consistently observed in all simulations, conducted under various setups. On the other hand, there is a large gap between the MSE of carrier estimates and the CR bound, which indicates the joint estimation algorithm is not statistically efficient.

3.5 Conclusion

For wideband CDMA systems with short spreading codes, multiuser detection can be accomplished through blind parameter estimation that enables the reconstruction of the effective signature waveforms. A MUSIC-like approach has been presented in this chapter for multiuser channel estimation. The algorithm, when combined with polynomial matrix rooting, can handle joint channel and frequency estimation if

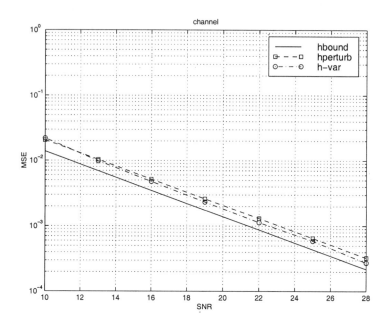

Figure 3.8 Performance of channel estimation.

the effect of carrier offset is too severe to ignore. Relative to the direct (and adaptive) receivers described in Chapter 2, these indirect, fully parametric approaches are better in performance but less favorable in implementation.

References

[1] R. O. Schmidt, "Multiple emitter location and signal parameter estimation," In *Proc. RADC Spectral Estimation Workshop*, pages 243–258, Griffiss AFB, NY, 1979.

[2] H. Liu and G. Xu, "A subspace method for signature waveform estimation in synchronous CDMA systems," *IEEE Trans. on Communications*, **COM-44**(10):1346–1354, October 1996.

[3] S. M. Kay, *Fundamentals of Statistical Signal Processing*, Englewood Cliffs, NJ: Prentice Hall, 1993.

[4] A. Franchi, C. Elia, and E. Colzi, "Maximum likelihood multipath

channel estimation for synchronous CDMA systems," In *Proc. Globecom*, pages 88–92, 1994.

[5] J. R. Treichler and B. G. Agee, "A new approach to multipath correction of constant modulus signals," *IEEE Trans. ASSP*, **31**(2):459–472, April 1983.

[6] S. Talwar, M. Viberg, and A. Paulraj, "Blind estimation of multiple cochannel digital signals using an antenna array," *IEEE Signal Processing Letters*, **1**(2):29–31, February 1994.

[7] D. Yellin and B. Porat, "Blind identification of FIR systems excited by discrete-alphabet inputs," *IEEE Trans. Signal Processing*, **SP-41**(3):1331–1339, 1993.

[8] G. Golub and C. Van Loan, *Matrix Computations*, second edition, Baltimore, MD: Johns Hopkins University Press, 1984.

[9] G. Xu and T. Kailath, "Fast subspace decomposition," *IEEE Trans. on Signal Processing*, **42**(3):539–551, March 1994.

[10] R. Lupas and S. Verdú, "Linear multiuser detectors for synchronous CDMA channels," *IEEE Trans. on Information Theory*, **35**(1):123–136, January 1989.

[11] U. Madhow and M. L. Honig, "MMSE interference suppression for direct-sequence spread sprectrum CDMA," *IEEE Trans. on Communications*, **42**(12):3178–3188, December 1994.

[12] J. H. Winters, J. Salz, and R. D. Gitlin, "The impact of antenna diversity on the capacity of wireless communication systems," *IEEE Trans. on Communications*, **42**(2/3/4):1740–1751, February/March/April 1994.

[13] B. Suard, G. Xu, H. Liu, and T. Kailath, "Channel capacity of spatial-division multiple-access schemes," In *Proc. 28th Asilomar Conference on Signals, Systems, and Computers*, pages 1159–1163, Pacific Grove, CA, November 1994.

[14] A. F. Naguid and A. Paulraj, "A base-station antenna array receiver for cellular DS/CDMA with M-ary orthogonal modulation," In *Proc. 28th Asilomar Conf. on Signals, Systems, and Computers*, pages 858–862, Pacific Grove, CA, November 1994.

[15] Fu Li, H. Liu, and R. J. Vaccaro, "Performance analysis for DOA estimation algorithms: Further unification, simplification, and observations," *IEEE Trans. Aerosp., Electron. Syst.*, **AES-**

29(3):1170–1184, October 1993.

[16] U. Mitra and H. V. Poor, "Adaptive receiver algorithm for near-far resistant CDMA," *IEEE Trans. on Communications*, **43**(4):1713–1724, April 1995.

[17] U. Madhow and M. L. Honig, "MMSE interference suppression for direct-sequence spread spectrum CDMA," *IEEE Trans. Communications*, **COM-42**(12):3178–3188, December 1994.

[18] R. Lupas and S. Verdú, "Linear multiuser detectors for synchronous code-division multiple-access channels," *IEEE Trans. on Information Theory*, **IT-35**(1):123–136, January 1989.

[19] Z. Xie, R. T. Short, and C. K. Rushforth, "A family of suboptimum detectors for coherent multiuser communications," *IEEE J. Selected Areas in Communications*, pages 683–690, May 1990.

[20] Z. Zvonar and D. Brady, "Suboptimum multiuser detector for synchronous CDMA frequency-selective Rayleigh fading channels," In *Globecom Mini-Conference on Communications Theory*, pages 82–86, 1992.

[21] A. J. Barabell, "Improving the resolution performance of eigenstructure-based direction-finding algorithm," In *Proc. IEEE ICASSP'83*, pages 336–339, August 1983.

[22] P. Stoica and A. Nehorai, "MUSIC, Maximum Likelihood and Cramér-Rao Bound," *IEEE Trans. on Acoustics, Speech, and Signal Processing*, **ASSP-37**(5):720–741, May 1989.

[23] P. P. Vaidyanathan, *Multirate Systems and Filter Banks*, Englewood Cliffs, NJ: Prentice Hall, 1993.

[24] P. P. Vaidyanathan, "Role of Anticausal Inverse in Multirate Filter-banks – Part I: System-Theoretic Fundamentals," *IEEE Trans. on Signal Processing*, **SP-43**(5):1090–1102, May 1995.

Appendix 3A: The Existence and Calculation of $\mathbf{G}(z)$

Before proceeding, recall the definition of the Smith form of a polynomial matrix [23].

CDMA With Short Codes: Indirect Approaches   77

Definition 1 *Given a $p \times r$ polynomial matrix $\mathbf{F}(z)$, it can always be expressed in Smith form $\mathbf{F}(z) = \mathbf{U}(z)\mathbf{S}(z)\mathbf{V}(z)$, where:*

1. *$\mathbf{U}(z)$, $\mathbf{V}(z)$ are unimodular polynomial matrices in z (i.e., with constant determinant).*

2. *$\mathbf{S}(z)$ is a $p \times r$ diagonal matrix with first ρ diagonal elements $\lambda_i(z)$, $0 \le i < \rho$ that are polynomials in z. The remaining diagonal elements of $\mathbf{S}(z)$ are zero. Here, ρ is the rank of $\mathbf{F}(z)$. The elements $\lambda_i(z)$ are given by $\lambda_i(z) = \frac{\Delta_{i+1}(z)}{\Delta_i(z)}$, $\Delta_0(z) = 1$, and $\Delta_i(z)$ is the greatest common divisor of all the $i \times i$ minors of $\mathbf{F}(z)$. $\lambda_i(z)$ is a factor of $\lambda_{i+1}(z)$.*

Theorem 1 [24] *For a $p \times r$ polynomial matrix $\mathbf{F}(z)$, there exists an FIR inverse $\mathbf{G}(z)$, which satisfies (3.32) if and only if the diagonal elements of $\mathbf{F}(z)$'s Smith form are $\lambda_i(z) = z^{-n_i}$, $0 \le i < r$.*

In our case, $p=L$, $r=L-1$. The above theorem asserts that if all of the $(L-1) \times (L-1)$ minors of $\mathbf{F}(z)$ are coprime with the exception of factor z^{n_i}, which holds almost surely in practice, there exists an FIR inverse $\mathbf{G}(z)$ that satisfies (3.32). We now calculate $\mathbf{G}(z)$ under the assumption that such an inverse exists.

Rewrite $\mathbf{F}(z)$ and $\mathbf{G}(z)$ in matrix polynomial form:

$$\mathbf{F}(z) = \sum_{i=0}^{K-1} \mathbf{F}(i) z^{-i}, \quad \mathbf{G}(z) = \sum_{i=0}^{B-1} \mathbf{G}(i) z^{-i} \qquad (3.45)$$

The time-domain equivalent of (3.32) is given by:

$$\underbrace{\begin{bmatrix} \mathbf{G}(0) & \mathbf{G}(1) & \cdots & \mathbf{G}(B-1) \end{bmatrix}}_{\mathbf{G_G}} \times$$

$$\underbrace{\begin{bmatrix} \mathbf{F}(0) & \mathbf{F}(1) & \cdots & \mathbf{F}(K) & 0 & \cdots & 0 \\ 0 & \mathbf{F}(0) & \cdots & \mathbf{F}(K-1) & \mathbf{F}(K) & \cdots & 0 \\ \vdots & & \ddots & \vdots & & \ddots & \vdots \\ 0 & 0 & \cdots & \mathbf{F}(0) & \cdots & \cdots & \mathbf{F}(K) \end{bmatrix}}_{\mathbf{F_F},\ [BL \times (K+B-1)(L-1)]} \qquad (3.46)$$

$$= \underbrace{\begin{bmatrix} 0 & \cdots & \mathbf{I} & \cdots & 0 \end{bmatrix}}_{\mathbf{I_o}}$$

where $\mathbf{I}_o$ has $\mathbf{I}_{L-1 \times L-1}$ at its n_0th block.

By choosing appropriate B, the length of $\mathbf{G}(z)$, and the delay index n_0, $\mathbf{G_G}$ can be determined as the solution of (3.46). To calculate $\mathbf{G}(z)$, we first choose B satisfying $B > (K-1)(L-1)$ so that the block Hankel matrix $\mathbf{F_F}$ constructed from $\mathbf{F}(z)$ is a tall matrix. Next, we identify the delay n_0 such that the n_0th diagonal block of the product $\mathbf{F_F^\dagger F_F}$ is $\mathbf{I}$. With n_0 so chosen, $\mathbf{G_G}$ can be calculated as:

$$\mathbf{G_G} = \mathbf{I}_o \mathbf{F_F^\dagger} \tag{3.47}$$

$\mathbf{G}(z)$ can be easily reconstructed from $\mathbf{G_G}$.

Appendix 3B: First and Second Order Derivatives of the $J(\phi, \mathbf{h}, \mathbf{U}_o)$

$$\frac{\partial J(\phi, \mathbf{h}, \hat{\mathbf{U}}_o)}{\partial \phi} = \mathbf{h}^H \mathbf{C}^H \mathbf{Z}_z'^H \hat{\mathbf{U}}_o \hat{\mathbf{U}}_o^H \mathbf{Z} \mathbf{C} \mathbf{h} + \mathbf{h}^H \mathbf{C}^H \mathbf{Z}^H \hat{\mathbf{U}}_o \hat{\mathbf{U}}_o^H \mathbf{Z}_z' \mathbf{C} \mathbf{h}$$

$$\approx \mathbf{h}^H \mathbf{C}^H \mathbf{Z}_z'^H \mathbf{U}_o \Delta \mathbf{U}_o^H \mathbf{Z} \mathbf{C} \mathbf{h} + \mathbf{h}^H \mathbf{C}^H \mathbf{Z}^H \Delta \mathbf{U}_o \mathbf{U}_o^H \mathbf{Z}_z' \mathbf{C} \mathbf{h}$$

$$\frac{\partial J(\phi, \mathbf{h}, \hat{\mathbf{U}}_o)}{\partial \bar{\mathbf{h}}_o} = \mathbf{h}^H \mathbf{C}^H \mathbf{Z}^H \hat{\mathbf{U}}_o \hat{\mathbf{U}}_o^H \mathbf{Z} \check{\mathbf{C}} + \mathbf{h}^T \mathbf{C}^T \mathbf{Z}^T \hat{\mathbf{U}}_o^* \hat{\mathbf{U}}_o^T \mathbf{Z}^* \check{\mathbf{C}}^*$$

$$\approx \mathbf{h}^H \mathbf{C}^H \mathbf{Z}^H \Delta \mathbf{U}_o \mathbf{U}_o^H \mathbf{Z} \check{\mathbf{C}} + \mathbf{h}^T \mathbf{C}^T \mathbf{Z}^T \Delta \mathbf{U}_o^* \mathbf{U}_o^T \mathbf{Z}^* \check{\mathbf{C}}^*$$

$$\frac{\partial J(\phi, \mathbf{h}, \hat{\mathbf{U}}_o)}{\partial \tilde{\mathbf{h}}_o} = j \mathbf{h}^H \mathbf{C}^H \mathbf{Z}^H \hat{\mathbf{U}}_o \hat{\mathbf{U}}_o^H \mathbf{Z} \check{\mathbf{C}} - j \mathbf{h}^T \mathbf{C}^T \mathbf{Z}^T \hat{\mathbf{U}}_o^* \hat{\mathbf{U}}_o^T \mathbf{Z}^* \check{\mathbf{C}}^*$$

$$\approx j \mathbf{h}^H \mathbf{C}^H \mathbf{Z}^H \Delta \mathbf{U}_o \mathbf{U}_o^H \mathbf{Z} \check{\mathbf{C}} - j \mathbf{h}^T \mathbf{C}^T \mathbf{Z}^T \Delta \mathbf{U}_o^* \mathbf{U}_o^T \mathbf{Z}^* \check{\mathbf{C}}^*$$

$$\frac{\partial^2 J(\phi, \mathbf{h}, \hat{\mathbf{U}}_o)}{\partial \phi^2} = \mathbf{h}^H \mathbf{C}^H \mathbf{Z}_z''^H \hat{\mathbf{U}}_o \hat{\mathbf{U}}_o^H \mathbf{Z} \mathbf{C} \mathbf{h} + 2 \mathbf{h}^H \mathbf{C}^H \mathbf{Z}_z'^H \hat{\mathbf{U}}_o \hat{\mathbf{U}}_o^H \mathbf{Z}' \mathbf{C} \mathbf{h}$$

$$+ \mathbf{h}^H \mathbf{C}^H \mathbf{Z}^H \hat{\mathbf{U}}_o \hat{\mathbf{U}}_o^H \mathbf{Z}_z'' \mathbf{C} \mathbf{h}$$

$$\frac{\partial^2 J(\phi, \mathbf{h}, \hat{\mathbf{U}}_o)}{\partial \phi \partial \bar{\mathbf{h}}_o} = \mathbf{h}^H \mathbf{C}^H \mathbf{Z}_z'^H \hat{\mathbf{U}}_o \hat{\mathbf{U}}_o^H \mathbf{Z} \check{\mathbf{C}} + \mathbf{h}^T \mathbf{C}^T \mathbf{Z}^T \hat{\mathbf{U}}_o^* \hat{\mathbf{U}}_o^T \mathbf{Z}'^* \check{\mathbf{C}}^*$$

$$+ \mathbf{h}^H \mathbf{C}^H \mathbf{Z}_z^H \hat{\mathbf{U}}_o \hat{\mathbf{U}}_o^H \mathbf{Z}' \check{\mathbf{C}} + \mathbf{h}^T \mathbf{C}^T \mathbf{Z}'^T \hat{\mathbf{U}}_o^* \hat{\mathbf{U}}_o^T \mathbf{Z}^* \check{\mathbf{C}}^*$$

$$\frac{\partial^2 J(\phi, \mathbf{h}, \hat{\mathbf{U}}_o)}{\partial \phi \partial \tilde{\mathbf{h}}_o} = j \mathbf{h}^H \mathbf{C}^H \mathbf{Z}_z'^H \hat{\mathbf{U}}_o \hat{\mathbf{U}}_o^H \mathbf{Z} \check{\mathbf{C}} - j \mathbf{h}^T \mathbf{C}^T \mathbf{Z}^T \hat{\mathbf{U}}_o^* \hat{\mathbf{U}}_o^T \mathbf{Z}'^* \check{\mathbf{C}}^*$$

$$j \mathbf{h}^H \mathbf{C}^H \mathbf{Z}_z^H \hat{\mathbf{U}}_o \hat{\mathbf{U}}_o^H \mathbf{Z}' \check{\mathbf{C}} - j \mathbf{h}^T \mathbf{C}^T \mathbf{Z}'^T \hat{\mathbf{U}}_o^* \hat{\mathbf{U}}_o^T \mathbf{Z}^* \check{\mathbf{C}}^*$$

$$\frac{\partial^2 J(\phi, \mathbf{h}, \hat{\mathbf{U}}_o)}{\partial \bar{\mathbf{h}}_o^2} = 2\operatorname{Re}[\check{\mathbf{C}}^H \mathbf{Z}^H \hat{\mathbf{U}}_o \hat{\mathbf{U}}_o^H \mathbf{Z} \check{\mathbf{C}}]$$

$$\frac{\partial^2 J(\phi, \mathbf{h}, \hat{\mathbf{U}}_o)}{\partial \bar{\mathbf{h}}_o \partial \tilde{\mathbf{h}}_o} = -2\operatorname{Im}[\check{\mathbf{C}}^H \mathbf{Z}^H \hat{\mathbf{U}}_o \hat{\mathbf{U}}_o^H \mathbf{Z} \check{\mathbf{C}}]$$

$$\frac{\partial^2 J(\phi, \mathbf{h}, \hat{\mathbf{U}}_o)}{\partial \tilde{\mathbf{h}}_o^2} = 2\operatorname{Re}[\check{\mathbf{C}}^H \mathbf{Z}^H \hat{\mathbf{U}}_o \hat{\mathbf{U}}_o^H \mathbf{Z} \check{\mathbf{C}}]$$

Note the first derivatives have already been expressed in terms of $\Delta \mathbf{U}_o^H$ by keeping only the first-order terms and using the fact that $\mathbf{U}_o^H \mathbf{Z} \mathbf{C} \mathbf{h} = \mathbf{0}$.

Chapter 4

CDMA With Long Codes: Space-Time Processing

4.1 Introduction

For CDMA with long spreading codes where joint detection is problematic, the primary way to counter fading is to combine multipath signals from the desired user using a RAKE receiver. Since introduced in [1], RAKE receivers that constructively exploit the multipath signals have been demonstrated to be very effective in alleviating multipath fading. When multiple antennas are available at the basestation, 2D RAKE receivers (i.e., space-time RAKE receivers) can be employed to profit from both the spatial and the temporal diversities [2].

Previous work on RAKE reception often assumes timing information of all multipath reflections [2]. The underlying idea is to enhance the signal strength by coherently combining all multipath reflections that stemmed from the desired user. A RAKE receiver thus has the same front structure as the optimum receiver depicted in Figure 1.10, except in RAKE signal detection is performed without accounting for outputs from other matched filters. A practical difficulty of this scheme is that in the presence of a large number of active users, multipath delay estimation is nontrivial [3]. Secondly, even if all multipath delays, $\{\tau_l\}$, are available at the basestation, signal processing on a particular multipath signal is prohibitive due to the resolution constraint of discrete samples. In addition to these limitations, the matched filter-based receiver may not be the most efficient way to use RAKE structure in the first place. Many other ways of combining (e.g., the frequency-domain approach by Zoltowski et al. [4]) are more effective in simultaneously incorporating the spatial and temporal diversities.

In the first half of this chapter, we will discuss the possibility of performing uplink 2D RAKE reception without requiring detailed knowledge of the multipath channels. The continuous-time data model of an antenna array CDMA system given in (1.14) is reformulated into discrete-time in Section 4.2. We then introduce the canonical structure of a 2D RAKE receiver and evaluate its potential for a general class of multipath channels. The optimum performance a 2D RAKE receiver can offer is derived in Section 4.2.1. In Section 4.2.2, we introduce a principal component (PC) method to estimate the minimum mean-square error RAKE receiver directly from the antenna outputs. The new approach outperforms traditional 2D RAKE schemes without imposing any constraint on the multipath structure. The data efficiency of the estimation is further improved in Section 4.2.3, where a deterministic least-squares (LS) algorithm is developed.

The second half of this chapter deals with blind detection in downlink CDMA communications. Compared with uplink, the downlink problem is quite different since each user only needs to extract its own signal from mixed CDMA signals broadcast from the base station. The commonly used downlink receiver consists of a linear channel equalizer followed by a despreader [5, 6]. The decoupling of equalization and despreading allows the receiver to handle IS-95 types of CDMA signals with long pseudorandom mask codes. In most applications, the channel equalizer has to be trained using a pilot sequence in a fading environment.

We will present a low-complexity downlink receiver capable of suppressing MAI and ISI without using training sequences. The development of the blind receiver is based on the following simple observations:

1. Downlink signals are perfectly synchronized at the transmitter.

2. All downlink signals go through the same multipath channel.

3. In synchronous transmission, users' long spreading codes can be chosen to be orthogonal at the symbol level.

In the absence of noise, the new blind receiver reduces to a zero-forcing (ZF) equalizer under mild conditions. Even with noise, our studies show that its performance approaches that of the ideal ZF receiver. Adaptive implementation is investigated in Section 4.3.2. We will adopt the simple gradient descent strategy and analyze the global convergence properties of the adaptive algorithm [7–10]. The step-size upper-bound is derived to ensure that the adaptive receiver converges to the global minimum.

4.2 Uplink

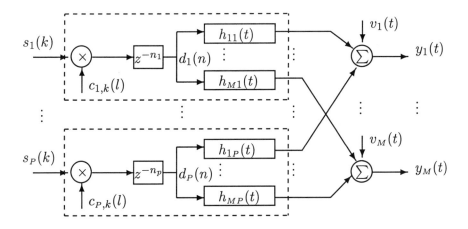

Figure 4.1 Antenna array CDMA with aperiodic spreading sequences.

Consider the antenna array CDMA system described in (1.13). In the presence of noise, the baseband signal received at the mth receiver is given by:

$$y_m(t) = \sum_{i=1}^{P} \sum_{n=-\infty}^{\infty} x_i(n) f_{mi}(t - nT) + v_m(t) \qquad (4.1)$$

The additive noise $\{v_m(t)\}$ is assumed to be of power σ_n^2 and is independent to the signals.

To facilitate our analysis, we shall cast the above CDMA system into a multi-input multi-output (MIMO) framework of Figure 4.1, with $\{d_i(n)\}$ denoting the spread chip sequences and $\{h_{mi}\}$ the composite channel responses of maximum length $L_c T_c$. The antenna outputs can be expressed in vector convolution form below:

$$\mathbf{y}(t) = [y_1(t) \cdots y_M(t)]^T = \sum_{i=1}^{P} \sum_{n=-\infty}^{\infty} d_i(n) \mathbf{h}_i(t - nT_c) + \mathbf{v}(t) \quad (4.2)$$

where $\mathbf{h}_i(t) = [h_{1i}(t) \cdots h_{Mi}(t)]^T$ and $\mathbf{v}(t) = [v_1(t) \cdots v_M(t)]^T$ represent the noise vector.

Note the modulated chip sequence from the ith user:

$$d_i(n) = s_i(k) c_{i,k}(n - kL - n_i), \quad k = \lfloor \frac{n - n_i}{L} \rfloor \quad (4.3)$$

The chip delay index n_i, $0 \leq n_i < L$, accounts for the relative discrete-time offset between users. In practice, this offset is usually known to the receiver after coarse synchronization during the initial access. The exact timing (within a fraction of a chip duration) of the direct-path signal, incorporated in $\mathbf{h}_i(t)$, is assumed to be unknown.

Sampling each antenna output at the chip rate yields:

$$\mathbf{y}(n) = \sum_{i=1}^{P} \sum_{l=1}^{L_c} \mathbf{h}_i(l) d_i(n - l + 1) + \mathbf{v}(n) \quad (4.4)$$

The general problem addressed here is the design and estimation of linear FIR receivers for the recovery of the transmitted signals, $\{s_i(n)\}$, from $\mathbf{y}(n)$.

It is worth pointing out that although the above expression is formulated in the context of antenna array systems, spatial oversampling is not the only way to create multiple outputs. By stacking consecutive samples within a chip in vector form (temporal oversampling), single antenna outputs can be cast into the same framework as in (4.4). The multiple output structure can also be made available using a combination of both spatial and temporal oversampling. In this case, the effective number of channels is $M = $ (*number of antennas*) × (*temporal oversampling rate*) [11].

4.2.1 Space-Time Receivers

The structure of the discrete-time 2D RAKE receiver is depicted in Figure 4.2. What we are concerned with now is the problem of receiver design (i.e., the selection of the filter coefficients) without the use of training sequences.

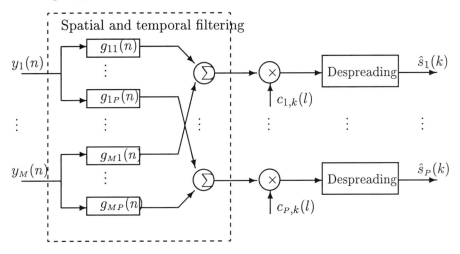

Figure 4.2 2D RAKE receivers.

Before proceeding, we shall first describe the exact operation of the 2D RAKE receiver. As seen, a vector FIR filter/equalizer $\mathbf{g}_i(n) = [g_{1,1}(n) \cdots g_{M,1}(n)]^T$ of order K is designated for each user to capture the diversities offered by the 2D channels. The filter output is defined as:

$$\tilde{d}_i(n-K) \stackrel{\text{def}}{=} \mathbf{g}_i(n) \otimes \mathbf{y}(n) = \sum_{k=1}^{K} \mathbf{g}_i^H(k)\mathbf{y}(n-k+1) \quad (4.5)$$

where the K-chip delay in $\tilde{d}_i(n-K)$ is introduced to accommodate multipath with maximum delay.

The filter output is despread with regenerated PN sequences to yield a symbol estimate, $\hat{s}_i(k)$. Upon defining:

$$\mathbf{g}_i = [\mathbf{g}_i^H(K) \ \mathbf{g}_i^H(K-1) \ \cdots \ \mathbf{g}_i^H(1)]^H \quad (4.6)$$
$$\mathbf{y}_K(n) = [\mathbf{y}^H(n-K+1) \ \mathbf{y}^H(n-K+2) \cdots \mathbf{y}^H(n)]^H \quad (4.7)$$

the whole reception process can be expressed as:

$$\begin{aligned}\hat{s}_i(k) &= \sum_{l=1}^{L} \tilde{d}_i(kL+n_i+l)c_{i,k}(l) \\ &= \sum_{l=1}^{L} \left(\mathbf{g}_i^H \mathbf{y}_K(kL+n_i+K+l)\right) c_{i,k}(l)\end{aligned} \quad (4.8)$$

Alternatively, the above processing can be realized by reversing the order of filtering and despreading. More specifically, one may first despread $\mathbf{y}(n)$ at different delays starting at $n = n_i, n_i+1, \cdots, n_i+K-1$:

$$\begin{aligned}\mathbf{z}_{i,K}(k) &\stackrel{\text{def}}{=} \begin{bmatrix} \sum_{l=1}^{L} \mathbf{y}(kL+n_i+l-1)c_{i,k}(l) \\ \vdots \\ \sum_{l=1}^{L} \mathbf{y}(kL+n_i+K+l-1)c_{i,k}(l) \end{bmatrix} \\ &= \mathbf{Y}_i(k)\mathbf{c}_i(k)\end{aligned} \quad (4.9)$$

where $\mathbf{Y}_i(k) = \begin{bmatrix} \mathbf{y}_K(kL+n_i+K) & \cdots & \mathbf{y}_K(kL+n_i+K+L) \end{bmatrix}$. The equalizer/combiner $\{\mathbf{g}_i(k)\}_{k=1}^{K}$ is then applied to the despread sequence $\mathbf{z}_{i,K}$ as follows:

$$\hat{s}_i(k) = \mathbf{g}_i^H \mathbf{z}_{i,K}(k) = \sum_{l=1}^{L} \tilde{d}_i(kL+n_i+l)c_{i,k}(l)$$

Comparing the above result with (4.8), the two schemes are clearly mathematically identical. The coefficients, $\{\mathbf{g}_i(k)\}_{i=1}^{P}$, determine the efficacy of the 2D RAKE receivers.

Coherent Combiners

The simplest filter used in 2D RAKE reception is the coherent combiner that matches the multipath channel. Assuming knowledge of the channel coefficients, signals from the desired user are coherently combined based on the channel characteristics as in a matched filter. For obvious reasons, the length of the receiver, K, is set to be equal to the channel length, L_c, and the receiver coefficients are simply chosen as:

$$\mathbf{g}_i(l) = \mathbf{h}_i(l), \quad l = 1, \cdots, L_c$$

In the remainder of this chapter, we shall refer to this scheme as the "coherent combiner." The coherent combiner is attractive in practice because of its simplicity. However, it does not account for the statistics of the MAI plus noise and therefore has some well-known disadvantages. In addition, it may also suffer from severe performance degradation if the delays of multipath signals are not far apart.

Optimum 2D RAKE Receivers

The discrete framework in (4.5) renders a well-posed linear FIR filter design problem, allowing us to benefit from the extensive work that has been done in the context of vector channel equalization [12–16].

Using the mean-square error (MSE) as the performance measure, the 2D RAKE receiver of order K that minimizes:

$$E\{|\mathbf{g}_i^H \mathbf{y}_K(n) - d_i(n-K)|^2\} \tag{4.10}$$

is defined as the MMSE 2D RAKE receiver. With straightforward manipulation, it is not difficult to derive that:

$$\mathbf{g}_{i,\text{MMSE}} = \mathbf{R}_{\mathbf{y}_K,\mathbf{y}_K}^{-1} \mathbf{r}_{i,K}$$

where $\mathbf{R}_{\mathbf{y}_K,\mathbf{y}_K} = E\{\mathbf{y}_K(n)\mathbf{y}_K^H(n)\}$ and $\mathbf{r}_{i,K} = E\{\mathbf{y}_K(n)d_i(n-K)\}$. To obtain an explicit expression, rewrite $\mathbf{y}_K(n)$ as:

$$\mathbf{y}_K(n) = \sum_{i=1}^{P} \mathbf{H}_i \mathbf{d}_{i,K}(n) + \mathbf{v}_K(n) \tag{4.11}$$

where:

$$\mathbf{H}_i \stackrel{\text{def}}{=} \underbrace{\begin{bmatrix} \mathbf{h}_i(L_c) & \cdots & \mathbf{h}_i(1) & 0 & \cdots & 0 \\ 0 & \mathbf{h}_i(L_c) & \cdots & \mathbf{h}_i(1) & \cdots & 0 \\ \vdots & \ddots & \ddots & \ddots & \ddots & \vdots \\ 0 & 0 & \mathbf{h}_i(L_c) & \mathbf{h}_i(L_c-1) & \cdots & \mathbf{h}_i(1) \end{bmatrix}}_{K+L_c \text{ blocks}}$$

$$\mathbf{d}_{i,K}(n) \stackrel{\text{def}}{=} [d_i(n-L_c-K-1) \cdots d_i(n-K) \cdots d_i(n)]^T \tag{4.12}$$

$$\mathbf{v}_K(n) \stackrel{\text{def}}{=} [\mathbf{v}_i^H(n-K-1)\ \mathbf{v}_i^H(n-K+1) \cdots \mathbf{v}_i^H(n)]^H \tag{4.13}$$

Proposition 1 *The vector equalizer that minimizes the MSE in (4.10) is given by:*

$$\mathbf{g}_{i,\text{MMSE}} = \mathbf{R}_{\mathbf{y}_K,\mathbf{y}_K}^{-1} \mathbf{r}_{i,K} \qquad (4.14)$$

where:

$$\mathbf{R}_{\mathbf{y}_K,\mathbf{y}_K} = \sum_{i=1}^{P} \mathbf{H}_i \mathbf{H}_i^H + \sigma_n^2 \mathbf{I},$$

$$\mathbf{r}_{i,K} = \mathbf{H}_i \mathbf{e}_L = [\mathbf{h}_i^H(1) \cdots \mathbf{h}_i^H(L_c), \mathbf{0} \cdots \mathbf{0}]^H, \qquad (4.15)$$

and $\mathbf{e}_l \stackrel{\text{def}}{=} [0 \cdots 0 \ 1 \ 0 \cdots 0]^T$ *with 1 at the lth position.*

The vector equalizer defined above offers the optimum performance among all linear FIR receivers. Unlike other 2D RAKE receivers that apply separate signal processing in time and spatial domains, the MMSE 2D RAKE receiver takes advantage of all possible diversities within the sample signals in a joint fashion. Such simultaneous exploitation has been demonstrated to be instrumental to ICI and MAI suppression in a multichannel setup [11, 14, 15].

Equivalence to Symbol Sequence Optimization

Toward this stage, our derivation has been based on the minimization of the MSE of the desired chip sequence rather than the symbol sequence. From a reception viewpoint, it is apparently more plausible to adopt the MSE of symbol estimates, $E\{|\hat{s}_i(n) - s_i(n)|^2\}$, as the optimization criterion. In this subsection, we show that the MMSE equalizer given in (4.14) also minimizes the MSE of symbol estimates. In other words, chip level optimization is equivalent to symbol lever optimization under the 2D RAKE structure (e.g., filtering followed up despreading).

The MMSE equalizer for symbol estimates is found by minimizing:

$$E\{|\mathbf{g}_i \mathbf{z}_{i,K}(k) - s_i(k)|^2\} = E\{|\mathbf{g}_i \mathbf{Y}_i(k)\mathbf{c}_i(k) - s_i(k)|^2\} \qquad (4.16)$$

with respect to $\mathbf{g}_{i,\text{MMSE-symbol}}$. Taking the derivative of the above expression and setting it to zero, we have:

$$\mathbf{g}_i = \mathbf{R}_{\mathbf{z}_{i,K},\mathbf{z}_{i,K}}^{-1} \mathbf{r}_{i,K} = \frac{1}{L} \left(L \mathbf{r}_{i,K} \mathbf{r}_{i,K}^H + \mathbf{R}_{\mathbf{p},\mathbf{p}} \right)^{-1} \mathbf{r}_{i,K} \qquad (4.17)$$

CDMA With Long Codes: Space-Time Processing

Recall from (4.14): that

$$g_{i,\text{MMSE}} = \mathbf{R}_{\mathbf{y}_K,\mathbf{y}_K}^{-1}\mathbf{r}_{i,K} = \left(\mathbf{r}_i\mathbf{r}_i^H + \mathbf{R}_{\mathbf{p},\mathbf{p}}\right)^{-1}\mathbf{r}_{i,K}$$

The following easily-proved lemma establishes the correspondence between $g_{i,\text{MMSE}}$ and $g_{i,\text{MMSE-symbol}}$ in the above equation.

Lemma 2 *Given an invertible matrix $\mathbf{R}$ and a column vector $\mathbf{r}$,*

$$(\mathbf{R}+\mathbf{r}\mathbf{r}^H)^{-1}\mathbf{r} = \alpha\mathbf{R}^{-1}\mathbf{r}$$

where $\alpha = 1 + \mathbf{r}^H\mathbf{R}^{-1}\mathbf{r}$.

Proof:
Using the matrix inversion lemma,

$$\begin{aligned}
(\mathbf{R}+\mathbf{r}\mathbf{r}^H)^{-1}\mathbf{r} &= \left(\mathbf{R}^{-1} - \frac{\mathbf{R}^{-1}\mathbf{r}\mathbf{r}^H\mathbf{R}^{-1}}{1+\mathbf{r}^H\mathbf{R}^{-1}\mathbf{r}}\right)\mathbf{r} \\
&= \mathbf{R}^{-1}\mathbf{r} - \frac{(\mathbf{r}^H\mathbf{R}^{-1}\mathbf{r})\mathbf{R}^{-1}\mathbf{r}}{1+\mathbf{r}^H\mathbf{R}^{-1}\mathbf{r}} \\
&= \frac{\mathbf{R}^{-1}\mathbf{r}}{1+\mathbf{r}^H\mathbf{R}^{-1}\mathbf{r}} \quad (4.18)
\end{aligned}$$

Therefore, $g_{i,\text{MMSE}}$ and $g_{i,\text{MMSE-symbol}}$ in (4.17) are identical within a scalar multiple.

Proposition 2 *The equalizer that minimizes the MSE of chip sequence estimates also minimizes the MSE of symbol sequence estimates.*

It is straightforward to derive the following result.

Proposition 3 *The MMSE of the ith symbol sequence estimates using a 2D RAKE receiver of order K is given by:*

$$\text{MMSE}_s = \frac{1}{1+L\mathbf{r}_{i,K}^H\mathbf{R}_{\mathbf{p},\mathbf{p}}\mathbf{r}_{i,K}} \quad (4.19)$$

Performance Comparison

By dealing with the composite channel instead of individual multipath, we are able to evaluate the capability of 2D RAKE receivers without imposing any specific assumptions on the channel structure. For the purpose of comparison, we derive in the following the MSE expressions for the coherent combiner and the MMSE equalizer. The results will provide insight to these approaches in a frequency-selective fading environment.

To begin with, decompose $\mathbf{y}_K(n)$ as follows:

$$\mathbf{y}_K(n) = d_i(n - K)\mathbf{r}_{i,K} + \mathbf{p}(n) \tag{4.20}$$

where $d_i(n - K)$, as seen from (4.10), is the signal of interest. $\mathbf{p}(n)$ denotes the ICI and MAI, plus the additive noise that needs to be suppressed.

Proposition 4 *The MSEs corresponding to the coherent combiner and the MMSE equalizer are given by:*

$$\text{MSE}_{cc} = \frac{\mathbf{r}_{i,K}^H \mathbf{R}_{\mathbf{p},\mathbf{p}} \mathbf{r}_{i,K}}{\|\mathbf{r}_{i,K}\|^4 + \mathbf{r}_{i,K}^H \mathbf{R}_{\mathbf{p},\mathbf{p}} \mathbf{r}_{i,K}} \tag{4.21}$$

$$\text{MMSE} = \frac{1}{1 + \mathbf{r}_{i,K}^H \mathbf{R}_{\mathbf{p},\mathbf{p}}^{-1} \mathbf{r}_{i,K}} \tag{4.22}$$

where $\mathbf{R}_{\mathbf{p},\mathbf{p}} = E\{\mathbf{p}(n)\mathbf{p}^H(n)\}$.

The derivation is procedurally straightforward and thus is omitted here. The MMSE in (4.22) sets the performance bound for all 2D RAKE receivers.

4.2.2 Blind 2D Reception

We are now in a position to consider the estimation of the coefficients of the MMSE receiver from the multichannel outputs without knowing the inputs. Given the parametric model in (4.11) and a finite number of observations, one can, in principle, obtain channel- and noise-parameter estimates via maximum likelihood (ML) estimation and then construct the MMSE 2D RAKE receiver based on the

CDMA With Long Codes: Space-Time Processing

expression given in (4.14) [17]. However, receiver design through optimum channel estimation is computationally expensive and more importantly, susceptible to model mismatch [14, 15]. From a practical viewpoint, there is a need for robust methods that yield satisfactory performance with affordable cost.

The explicit expression for the MMSE equalizer derived in the previous section suggests that a direct estimation procedure is possible. Indeed, as seen from (4.14), in order to calculate $\mathbf{g}_{i,\text{MMSE}}$ one only needs to estimate $\mathbf{R}_{\mathbf{y}_K,\mathbf{y}_K}$, which obviously can be approximated with the data covariance matrix and the vector $\mathbf{r}_{i,K}$. In the following, we derive a principal component algorithm to determine $\mathbf{r}_{i,K}$ from $\{\mathbf{y}(n)\}$.

The Principal Component Algorithm

The foundation to the principal component method, initially formulated for array response vector estimation for CDMA systems in a frequency nonselective environment [18], is the observation that the pre- and post-despreading covariance matrices of antenna outputs differ only by a rank-one matrix defined by $\mathbf{r}_{i,K}$. Therefore, $\mathbf{r}_{i,K}$ can be estimated as the principal eigenvector of the difference matrix. Similar approaches have been suggested to estimate $\{\mathbf{a}_i(l)\}$ in (1.3) when the multipath channels are frequency-selective [2]. However, their performance hinges on the accurate estimation of multipath delays. By dealing with composite channels rather than individual multipath signals, no restrictive assumptions are required in the present discrete-time framework.

From (4.20):

$$\mathbf{R}_{\mathbf{y}_K,\mathbf{y}_K} = \mathbf{r}_{i,K}\mathbf{r}_{i,K}^H + \mathbf{R}_{\mathbf{p},\mathbf{p}} \tag{4.23}$$

Consequently, $\mathbf{R}_{\mathbf{Y}_i} = E\{\mathbf{Y}_i(k)\mathbf{Y}_i^H(k)\} = L\mathbf{R}_{\mathbf{y}_K,\mathbf{y}_K} = L\mathbf{r}_i\mathbf{r}_i^H + L\mathbf{R}_{\mathbf{p},\mathbf{p}}$.

By despreading the antenna outputs, the power of the desired signal is enhanced by a factor of L^2, whereas the power of interference increases linearly by a factor of L. Using (4.9) and (4.20),

$$\mathbf{z}_{i,K}(k) = \mathbf{Y}_i(k)\mathbf{c}_i(k) = Ls_i(k)\mathbf{r}_{i,K} + \sum_{l=1}^{L}\mathbf{p}(kL + n_i + K + l - 1)c_i(k,l)$$

Therefore:
$$\mathbf{R}_{\mathbf{z}_{i,K}\mathbf{z}_{i,K}} = E\{\mathbf{z}_{i,K}, \mathbf{z}_{i,K}^H\} = L^2 \mathbf{r}_{i,K} \mathbf{r}_{i,K}^H + L\mathbf{R}_{\mathbf{p},\mathbf{p}} \qquad (4.24)$$

Comparing the above covariance matrix with $\mathbf{R}_{\mathbf{Y}_i}$, it is readily seen that:
$$\mathbf{R}_{\mathbf{z}_{i,K},\mathbf{z}_{i,K}} - \mathbf{R}_{\mathbf{Y}_i} = L(L-1)\mathbf{r}_{i,K}\mathbf{r}_{i,K}^H \qquad (4.25)$$

The above result indicates that the span of $\mathbf{R}_{\mathbf{z}_{i,K},\mathbf{z}_{i,K}} - \mathbf{R}_{\mathbf{Y}_i}$ is defined by vector $\mathbf{r}_{i,K}$. Hence, $\mathbf{r}_{i,K}$ can be determined as the principal eigenvector of $\mathbf{R}_{\mathbf{z}_{i,K},\mathbf{z}_{i,K}} - \mathbf{R}_{\mathbf{Y}_i}$. In practice, $\mathbf{R}_{\mathbf{z}_i,\mathbf{z}_i} - \mathbf{R}_{\mathbf{Y}_i}$ is unknown and is therefore estimated by the sample correlation matrix $\hat{\mathbf{R}}_{\mathbf{z}_{i,K},\mathbf{z}_{i,K}} - \hat{\mathbf{R}}_{\mathbf{Y}_i}$:

$$\frac{1}{N}\left(\sum_{k=1}^{N} \mathbf{Y}_i(k)\mathbf{c}_i(k)\mathbf{c}_i^H(k)\mathbf{Y}_i^H(k) - \sum_{k=1}^{N} \mathbf{Y}_i(k)\mathbf{Y}_i^H(k)\right) \qquad (4.26)$$

Notice that since:
$$\mathbf{r}_{i,K} = [\mathbf{h}_i^H(1) \cdots \mathbf{h}_i^H(L_c), \mathbf{0} \cdots \mathbf{0}]^H$$

The above method also provides an effective way to estimate the multipath channel, $\{\mathbf{h}_i(l),\ l = 1, \cdots, L_c\}$.

Once we obtain the $\mathbf{R}_{\mathbf{y}_{i,K},\mathbf{y}_{i,K}}$ and $\mathbf{r}_{i,K}$ estimates, the MMSE equalizer can be computed accordingly. The following summarizes the estimation procedure:

1. Calculate the data covariance matrices as:

$$\hat{\mathbf{R}}_{\mathbf{Y}_i} = \frac{1}{N}\left(\sum_{k=1}^{N} \mathbf{Y}_i(k)\mathbf{Y}_i^H(k)\right) \qquad (4.27)$$

$$\hat{\mathbf{R}}_{\mathbf{z}_{i,K},\mathbf{z}_{i,K}} = \frac{1}{N}\left(\sum_{k=1}^{N} \mathbf{Y}_i(k)\mathbf{c}_i(k)\mathbf{c}_i^H(k)\mathbf{Y}_i^H(k)\right) \qquad (4.28)$$

2. Estimate $\hat{\mathbf{r}}_{i,K}$ as the principal eigenvector of $\hat{\mathbf{R}}_{\mathbf{Y}_i} - \hat{\mathbf{R}}_{\mathbf{z}_{i,K},\mathbf{z}_{i,K}}$.

3. The MMSE equalizer within a scalar multiple is given by:

$$\hat{\mathbf{g}}_{i,\text{MMSE}} = \hat{\mathbf{R}}_{\mathbf{Y}_i}^{-1}\hat{\mathbf{r}}_{i,K}, \quad i = 1, \cdots, P \qquad (4.29)$$

Theoretically, the above approach provides unbiased estimates of the MMSE equalizer given sufficient data samples. However, since $\mathbf{r}_{i,K}$ is estimated individually for each user, as opposed to joint ML estimation, the principal component method is not statistically efficient. Nevertheless, it possesses the robustness and simplicity essential to practical implementation. In the remainder of this book, we shall refer to the MMSE receiver obtained through the PC approach as the principal component (PC) MMSE receiver.

The PC algorithm is similar to the 2D RAKE receiver proposed by Zoltowski et al. in [19, 20]. The algorithm there is premised on despreading at a large number of delay times, on the order of L, the number of chips per bit, as opposed to K as in the PC MMSE algorithm proposed above. On the other hand, the algorithm is advantageous in performance if one samples faster than the chip rate. In this case the temporal characteristics of the other $P-1$ users and the receiver noise are generally altered by passing the signal through an FIR filter based on commensurate sampling of the spreading waveform of the ith user.

4.2.3 Examples

To illustrate the performance of the PC method, we present several numerical examples and compare the output SINR defined below for different approaches:

$$\text{SINR} = -10 \log_{10} \text{MSE} \quad [\text{dB}]$$

All examples here involve an 8-element antenna array and CDMA signals with a spreading factor of 32. Temporal oversampling is not used (i.e., sampling at the chip rate). The PN sequences are randomly generated rather than created using shift registers. Each composite channel comprises direct-path and multipath components – the delays and the number of multipath components were chosen uniformly from $[0 \ 3T_c]$ and $[1, 10]$, respectively. The maximum channel length is thus 3. The raised-cosine function with a roll-off factor of 0.5 is used and the input SNR is set at 20 dB. A total of 200 Monte Carlo runs are performed for each simulation. We assume mild power control in the

94 *Signal Processing Applications in CDMA Communications*

system so that the power variations among users are within 6 dB. The transmitted power of each user is defined as the total power received by the antenna array:

$$P_i = \sum_{l=1}^{L_c} \|\mathbf{h}_i(l)\|^2, \quad i = 1, \cdots, P \tag{4.30}$$

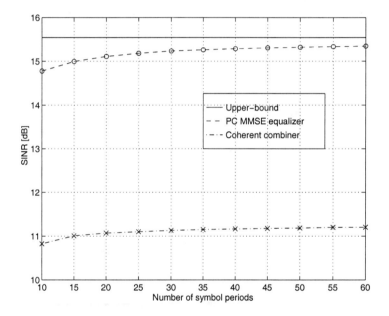

Figure 4.3 SINR versus number of samples.

Example 1: The first example involves 15 users. Data samples within $10 \sim 60$ symbol periods are used for MMSE receiver estimation. For comparison, the channel estimate obtained from $\hat{\mathbf{r}}_{i,K}$ is used in the coherent combiner. Figure 4.3 plots the output SINRs corresponding to the coherent combiner and the PC MMSE receiver for the first user. The superiority of the proposed receiver is clearly demonstrated. Note that as the number of samples increases, the performance of the PC method approaches the 2D RAKE receiver upper-bound. However, the noticeable gap between the upper-bound and the simulation SINRs implies that the PC estimator is not statistically efficient.

In Figure 4.4, the difference between the simulation SINRs and their theoretic upper-bound is plotted for all users in the systems. The top three lines, corresponding to the PC method using samples within 30, 45, and 60 symbol periods, are close to zero for all users. This demonstrates that the proposed estimation method is near-far resistant – a feature highly desirable in CDMA communications. At the same time, it is not surprising to see that the bottom three lines, corresponding to the coherent combiner, have large variations due to imperfect power control even though the channel estimates from the PC method are satisfactory.

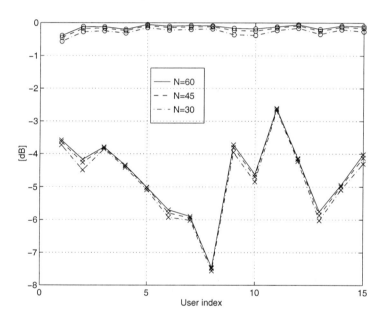

Figure 4.4 Simulation SINRs minus optimum SINRs.

Example 2: In this example, we fix N at 30 and vary the number of users to study how variations of MAI affect the output SINR of the desired user. The results are plotted in Figure 4.5. As expected, the SINRs decrease as the number of users increases. In addition, we observe that there is a consistent gap between the performance of the PC MMSE receiver and that of the coherent combiner.

The theoretic SINR of the MMSE receiver versus the filter length

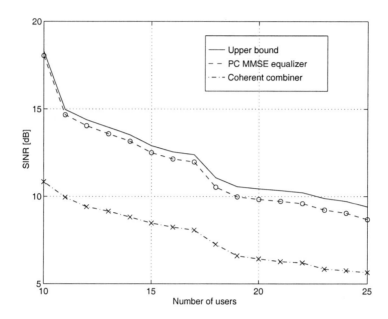

Figure 4.5 SINR versus number of users.

is plotted in Figure 4.6 for different P values. In all cases, a longer equalizer yields a higher SINR. However, the performance improvement due to an additional tap is evident only when K is small. After K reaches L_c, the output SINRs become saturated in most scenarios except for $P = 6$, where the SINR continues to increase until $K = 6$.

Data Efficient Deterministic Receiver

One of the limitations of the PC method is its reliance on the covariance matrices of antenna outputs. In a rapidly varying environment, the available data may not be adequate for data covariance matrix-based approaches to yield satisfactory parameter estimates. The finite data effect is inherent and thus difficult to overcome unless sufficient information about the underlying data structure is available.

Our goal in this section will be to find data-efficient "deterministic" solutions for the 2D RAKE receiver [11]. We are particularly interested in the zero-forcing receiver for underloaded CDMA systems. In this

CDMA With Long Codes: Space-Time Processing

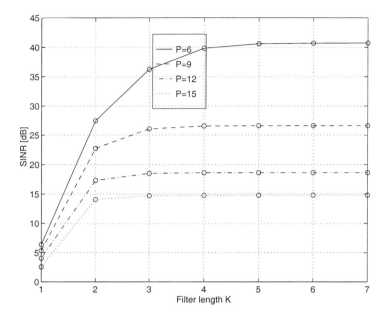

Figure 4.6 SINR versus filter length.

case, the number of effective channels is greater than the number of users, and it is feasible to perfectly recover the transmitted signals in the absence of noise (zero-forcing). The method presented here builds on an almost trivial observation, that is, knowledge of the PN spreading sequences can sidestep the use of training sequences and allows us to formulate a least-squares problem to determine the vector equalizer.

Ignore the noise term and rewrite (4.11) as:

$$\mathbf{y}_K(n) = \underbrace{\begin{bmatrix} \mathbf{H}_1 & \cdots & \mathbf{H}_P \end{bmatrix}}_{\stackrel{\text{def}}{=} \mathbf{H}} \begin{bmatrix} \mathbf{d}_{1,K}^H(n) \cdots \mathbf{d}_{P,K}^H(n) \end{bmatrix}^H$$

It is seen that the dimension of the channel matrix $\mathbf{H}$ is $MK \times P(L_c + K)$. Therefore, when $M > P$, there exists a K such that $\mathbf{H}$ has more rows than columns. The possibility of a zero-forcing receiver arises. All signals can be separated by applying the pseudoinverse of $\mathbf{H}$, $\mathbf{H}^\dagger$, to $\mathbf{y}_K(n)$:

$$\mathbf{H}^\dagger \mathbf{y}_K(n) = \mathbf{H}^\dagger \mathbf{H} \left[\mathbf{d}_{1,K}^H(n) \cdots \mathbf{d}_{P,K}^H(n)\right]^H = \left[\mathbf{d}_{1,K}^H(n) \cdots \mathbf{d}_{P,K}^H(n)\right]^H$$

For more discussions on zero-forcing equalization, see [15] and the references therein.

Note that without using the composite multipath channels, complete nulling of interference using spatial beamforming requires the number of receivers to be larger than the *total* number of multipath reflections [21].

To find the deterministic solution, assuming $\mathbf{g}_i$ is a zero-forcing equalizer for the ith user,

$$\mathbf{g}_i^H \mathbf{Y}_i(k) = [d_i(kL+n_i), \cdots, d_i(kL+n_i+L)] = s_i(k)\mathbf{c}_i^H(k) \quad (4.31)$$

The only unknowns in the above equation are $\mathbf{g}_i$ and $\{s_i(n)\}$. We may thus formulate a minimization problem with respect to $\mathbf{g}_i(n)$ and $s_i(1), s_i(2), \cdots, s_i(N)$ given a finite number of observations, $\mathbf{Y}_i(k)$, $k = 1, 2, \cdots, N$:

$$\arg \min_{\mathbf{g}, s_i(1), \cdots, s_i(N)} \sum_{k=1}^N \|\mathbf{g}^H \mathbf{Y}_i(k) - s_i(k)\mathbf{c}_i^H(k)\|^2 \quad (4.32)$$

Representing the above minimization problem in a matrix form:

$$\begin{bmatrix} \mathbf{Y}_i^H(1) & -\mathbf{c}_i(1) & 0 & \cdots & 0 \\ \mathbf{Y}_i^H(2) & 0 & -\mathbf{c}_i(2) & \cdots & 0 \\ \vdots & \vdots & 0 & \ddots & \vdots \\ \mathbf{Y}_i^H(N) & 0 & \cdots & 0 & -\mathbf{c}_i(N) \end{bmatrix} \begin{bmatrix} \mathbf{g}_i \\ s_i(1) \\ \vdots \\ s_i(N) \end{bmatrix} = 0 \quad (4.33)$$

It is seen that there are NL equations and $MK+N$ unknowns, which evidently defines an overdetermined least-squares problem when N is large enough. Therefore, the vector equalizer $\mathbf{g}_i$ can be determined from the nontrivial solution of (4.33). Since the parameters of interest are the coefficients in $\mathbf{g}_i$, structure and redundancy can be exploited to simplify the solution.

Proposition 5 *Given* $\{\mathbf{Y}_i(k)\}_{i=1}^N$, $\mathbf{g}_i$ *that satisfies (4.33) is found as the least significant eigenvector of* $\left(\hat{\mathbf{R}}_\mathbf{Y} - \hat{\mathbf{R}}_{\mathbf{z},\mathbf{z}}/L\right)$, *where* $\hat{\mathbf{R}}_\mathbf{Y}$ *and* $\hat{\mathbf{R}}_{\mathbf{z},\mathbf{z}}$ *are defined in (4.27) and (4.28), respectively.*

CDMA With Long Codes: Space-Time Processing

To verify that the above solution is indeed zero-forcing, decompose $\mathbf{Y}_i(k)$ into:

$$\mathbf{Y}_i(k) = s_i(k)\mathbf{r}_{i,K}\mathbf{c}_i^H(k) + \tilde{\mathbf{H}}\mathbf{P}(k)$$

where $\tilde{\mathbf{H}}$ is the $\mathbf{H}$ matrix without vector $\mathbf{r}_{i,K}$, and $\mathbf{P}(k)$ denotes the interfering signals. Notice that:

$$\begin{aligned}\mathbf{Y}_i(k)\mathbf{Y}_i(k) &= L\mathbf{r}_{i,K}\mathbf{r}_{i,K}^H + 2\Re\left(s_i(k)\mathbf{r}_{i,K}\mathbf{c}_i^H(k)\mathbf{P}(k)\tilde{\mathbf{H}}^H\right) \\ &+ \tilde{\mathbf{H}}\mathbf{P}(k)\mathbf{P}^H(k)\tilde{\mathbf{H}}^H \end{aligned} \quad (4.34)$$

Using

$$\mathbf{z}_{i,K}(k) = \mathbf{Y}_i(k)\mathbf{c}_i(k) = Ls_i(k)\mathbf{r}_{i,K} + \tilde{\mathbf{H}}\mathbf{P}(k)\mathbf{c}_i(k),$$

it is easily shown that:

$$\begin{aligned}\mathbf{z}_{i,K}(k)\mathbf{z}_{i,K}^H(k) &= L\left(L\mathbf{r}_{i,K}\mathbf{r}_{i,K}^H + 2\Re\left(s_i(k)\mathbf{r}_{i,K}\mathbf{c}_i^H\mathbf{P}(k)\tilde{\mathbf{H}}^H\right)\right) \\ &+ \tilde{\mathbf{H}}\mathbf{P}(k)\mathbf{c}_i(k)\mathbf{c}_i^H(k)\mathbf{P}^H(k)\tilde{\mathbf{H}}^H\end{aligned}$$

Consequently,

$$\hat{\mathbf{R}}_\mathbf{Y} - \hat{\mathbf{R}}_{\mathbf{z},\mathbf{z}}/L = \tilde{\mathbf{H}}\left(\sum_{k=1}^N \mathbf{P}(k)\mathbf{P}^H(k) - \sum_{k=1}^N \mathbf{P}(k)\mathbf{c}_i(k)\mathbf{c}_i^H(k)\mathbf{P}^H(k)\right)\tilde{\mathbf{H}}^H$$

The column span of the above matrix is defined by the interfering channel matrix, $\tilde{\mathbf{H}}$, which has at least one null vector under the assumption that $\mathbf{H}$ is of full column rank. An equalizer is zero-forcing if:

$$\mathbf{g}_i^H \mathbf{r}_{i,K} = \alpha \neq 0, \quad \mathbf{g}_i^H \tilde{\mathbf{H}} = \mathbf{0}$$

The least significant null vector of $\left(\hat{\mathbf{R}}_\mathbf{Y} - \hat{\mathbf{R}}_{\mathbf{z},\mathbf{z}}/L\right)$ is orthogonal to $\tilde{\mathbf{H}}$ but not to $\mathbf{r}_{i,K}$ since $\mathbf{r}_{i,K}$ is linearly independent of $\tilde{\mathbf{H}}$ and thus qualifies as a zero-forcing equalizer.

The deterministic method is also applicable to systems with $P > M$. Its performance, however, is difficult to predict without a full-scale performance analysis. The most suitable application of the least-squares method is probably in underloaded CDMA systems with high SNR. Otherwise, the PC MMSE equalizer that accounts for the statistics of the input signals is preferable, provided that a reasonable number of data samples are available.

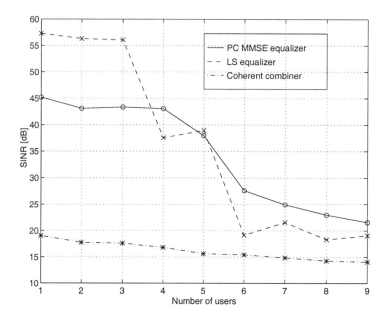

Figure 4.7 SINRs in an underloaded system.

Example 3: The last example serves to compare the performance of different approaches in underloaded systems with SNR at 40 dB. The filter length was fixed at $K = 3$. Data samples within 10 symbol periods were used in the least-squares method and the PC method. The performance comparison is plotted in Figure 4.7. When the number of users is small, the least-squares method shows its advantages in interference suppression. Note that with $K = 3$, one can in principle null out signals from 3 cochannel users. After the total number of users reaches 4, the zero-forcing condition no longer holds, and the performance of the least-squares approach drops below that of the PC MMSE receiver. It is interesting to notice that, unlike the coherent combiner and the PC MMSE receiver, the output SINR corresponding to the least-squares method does not decrease monotonically with the number of users. This is explainable considering the fact that $\mathbf{r}_{i,K}$ of the desired signal is not incorporated in the least-squares approach. It is possible that with an additional user, the resulting equalizer can better accommodate the desired signal and thus increases the SINR.

4.3 Downlink

In most existing CDMA systems, the same detection schemes are used in both uplink and downlink. Since downlink CDMA signals possess some intrinsic structures that are not available in uplink, we will discuss in the remainder of this chapter high-performance downlink signal reception by exploiting structural information.

In CDMA downlink, all users' signals are transmitted synchronously. By setting the relative delays to zero in (1.2), we obtain the expression for downlink CDMA signal as:

$$x(t) = \sum_{i=1}^{P} x_i(t) = \sum_{i=1}^{P} \sum_{k=-\infty}^{\infty} s_i(k) w_{i,k}(t - kT) \quad (4.35)$$

The time varying spreading waveform $w_{i,k}(t)$ is given in (1.5) of Chapter 1.

In downlink modulation, individual data sequences are first spread by orthogonal spreading codes (e.g., the Walsh codes). The spread sequences are then combined and overlaid with a long pseudorandom mask code. The lth chip within the kth symbol can thus be decomposed as:

$$c_{i,k}(l) = c_i(l) o_k(l), \quad i = 1, \cdots, P$$

where $c_i(l)$ is the short and periodic-spreading code for the ith user; and $o_k(l) \in \{\pm 1\}$ is the overlaid mask code that is common for all users. Note that the compound-spreading codes are still mutually orthogonal for each symbol period if $\{c_i(l)\}_{i=1}^{P}$ satisfy:

$$\sum_{l=1}^{L} c_i(l) c_j(l) = L \delta_{i,j} \quad (4.36)$$

Because of this inherent orthogonality, downlink signal reception is usually performed in two steps: (1) channel equalization that compensates the "common" multipath channel effect and restores the orthogonality, and (2) despreading that extracts the desired signal. The issue to be studied here is the design of channel equalizers.

For general purposes, we will consider a multireceiver downlink CDMA model depicted in Figure 4.8. M denotes the number of chan-

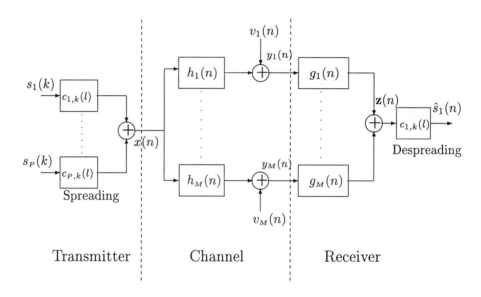

Figure 4.8 CDMA downlink over multiple FIR channels.

nels, obtained either through oversampling or multiple receiving antennas. The multipath channel effect is characterized by the single-input multi-output (SIMO) channel responses $\{h_m(n)\}$. The received signal through the mth channel is thus $y_m(n) = h_m(n) * x(n) + v_m(n)$.

Collecting chip-rate samples in vector form, the mth channel output can be expressed as:

$$\mathbf{y}_m(k) = \mathcal{T}_{L+l_h+l_t}(\mathbf{h}_m)\mathbf{x}(k) + \mathbf{v}_m(k) \qquad (4.37)$$

where:

$$\begin{aligned}
\mathbf{y}_m(k) &= [y_m(kL - l_h) \quad \cdots \quad y_m(kL + l_t)]^H \\
\mathbf{x}(k) &= [x(kL - l_h - L_c + 1) \quad \cdots \quad x(kL + l_t)]^H \\
\mathbf{v}_m(k) &= [v_m(kL - l_h) \quad \cdots \quad v_m(kL + l_t)]^H
\end{aligned}$$

and:

$$\mathcal{T}_{M+l_h+l_t}(\mathbf{h}_m) = \begin{bmatrix} h_m(L_c) & \cdots & h_m(1) & 0 & \cdots & 0 \\ 0 & h_m(L_c) & \cdots & h_m(1) & \cdots & 0 \\ \vdots & \vdots & \vdots & & \ddots & \vdots \\ 0 & \cdots & 0 & h_m(L_c) & \cdots & h_m(1) \end{bmatrix}$$

CDMA With Long Codes: Space-Time Processing

is the $(L+l_h+l_t)$-row Toeplitz convolutional matrix. l_h and l_t are the lengths of chip samples from the previous and the next symbol period.

4.3.1 Constrained Equalization

Let the first user be the user of interest. The downlink signal can be rewritten as:

$$\mathbf{x}(k) = \mathbf{x}_1(k) + \sum_{i=2}^{P} \mathbf{x}_i(k) \qquad (4.38)$$

where the second term stands for MAI. The desired user's signal $\mathbf{x}_1(n)$ can be further decomposed as:

$$\mathbf{x}_1(k) = \mathbf{c}_1(k)s_1(k) + \mathbf{c}_{1,t}(k-1)s_1(k-1) + \mathbf{c}_{1,h}(k-1)s_1(k+1) \qquad (4.39)$$

where:

$$\begin{aligned}
\mathbf{c}_1(k) &= [0 \cdots c_{1,k}(1) \cdots c_{1,k}(L) \cdots 0]^T \\
\mathbf{c}_{1,t}(k) &= [c_{1,k-1}(L - L_c - l_h + 1) \cdots c_{1,k-1}(L)\, 0 \cdots 0]^T \\
\mathbf{c}_{1,h}(k) &= [0 \cdots 0\, c_{1,k+1}(1) \cdots c_{1,k+1}(l_t)]^T
\end{aligned}$$

The last two terms in (4.39) are the ISI.

As shown in Figure 4.8, the outputs of the channel equalizers are first combined and then despread to form a symbol estimate. The input to the despreader is:

$$\mathbf{z}(k) = \sum_m \mathcal{T}(\mathbf{g}_m)\mathbf{y}_m(k) = \sum_m \mathcal{T}(\mathbf{y}_m(k))\mathbf{g}_m = \mathbf{Y}(k)\mathbf{g} \qquad (4.40)$$

where:

$$\begin{aligned}
\mathbf{Y}(k) &= [\mathcal{T}(\mathbf{y}_1(k)) \cdots \mathcal{T}(\mathbf{y}_M(k))] \\
\mathbf{g} &= [\mathbf{g}_1^H \cdots \mathbf{g}_M^H]^H
\end{aligned}$$

Since the despreading operation is fixed, the goal of the downlink receiver design is to find a set of equalizers $\{g_m(l)\}_{m=1}^{M}$ to suppress the MAI and ISI as much as possible.

Blind Equalization

In the absence of noise, a perfect channel equalizer would be the ZF equalizer. Its output satisfies:

$$f(l) = \sum_m h_m(l) * g_m(l) = \delta(l - n_0) \qquad (4.41)$$

where n_0 is an appropriate delay index. In many cases, it has been observed that selecting delay n_0 to be in the middle of equalized channel response results in good equalizer performance [22]. We choose $n_0 = (L_c + K - 1)/2$, where K is the length of the equalizer. Equation (4.41) tells us that in noiseless cases, downlink CDMA signals will preserve their orthogonality after passing a perfect channel equalizer. This suggests a means to design the channel equalizer $\{\mathbf{g}_i\}$.

In most practical CDMA systems, the number of active users is usually much less than the system spreading gain. Define:

$$\mathbf{c}_i^o(k) = [\; \underbrace{0 \cdots 0}_{(l_h + n_0) \text{ 0s}} \quad c_{i,k}(1) \quad \cdots \quad c_{i,k}(L) \quad \underbrace{0 \cdots 0}_{(l_t + K - n_0 - 1) \text{ 0s}} \;]^H$$

where $c_{i,k}(l)$, $i \in \{P+1, \cdots, L\}$ is any "unused" spreading code. Because of the default orthogonality, the following normal equation holds for a "perfect" equalizer:

$$\mathbf{c}_i^o(k)^H \mathbf{z}(k) = \mathbf{c}_i^o(k)^H \mathbf{Y}(k)\mathbf{g} = 0 \qquad (4.42)$$

Let $\mathbf{C}_o(k)$ be the matrix formed by column vectors $\mathbf{c}_i^o(k)$, $i \in \{P+1, \cdots, M\}$. Based on relation (4.42), a blind channel equalizer can be found by forcing the projection of its output onto unused orthogonal codes to be zero. More specifically, a "good" equalizer should be a solution that minimizes the output energy given below,

$$\mathbf{g}_{\text{pro}} = \arg\min_{\mathbf{g}} \frac{1}{N} \sum_{k=1}^{N} \|\mathbf{C}_o^H(k)\mathbf{Y}(k)\mathbf{g}\|^2 \qquad (4.43)$$

where N is the number of observation vectors.

The dimension of $\mathbf{C}_o(k)$ needs to be chosen properly to avoid solutions other than the ZF solution. For the time being, we use all possible $\mathbf{c}_l^o(k)$, that is:

$$\mathbf{C}_o(k) = [\mathbf{c}_{P+1}^o(k), \cdots, \mathbf{c}_M^o(k)]$$

(4.43) essentially minimizes an output energy function. To avoid trivial solutions, we introduce a constraint to prevent signal cancellation while minimizing the right side of (4.43). In particular, we force the projection of the equalizer output onto the desired user's spreading code to have fixed energy so that the desired user's signal is preserved.

To summarize, the described equalizer is a solution of the following constrained minimum output energy (C-MOE) problem:

$$\mathbf{g}_{pro} = \arg\min_{\mathbf{g}} \frac{1}{N} \sum_{k=1}^{N} \|\mathbf{C}_o^H(k)\mathbf{Y}(k)\mathbf{g}\|^2$$

$$\text{subject to } \frac{1}{N} \sum_{k=1}^{N} \|\mathbf{c}_1^o(k)^H \mathbf{Y}(k)\mathbf{g}\|^2 = 1 \quad (4.44)$$

Upon defining:

$$\mathbf{R}_o(N) = \frac{1}{N} \sum_{k=1}^{N} [\mathbf{Y}^H(k)\mathbf{C}_o(k)\mathbf{C}_o^H(k)\mathbf{Y}(k)]$$

$$\mathbf{R}_s(N) = \frac{1}{N} \sum_{k=1}^{N} [\mathbf{Y}^H(k)\mathbf{c}_1^o(k)\mathbf{c}_1^o(k)^H \mathbf{Y}(k)]$$

the minimization problem can be reformulated as:

$$\mathbf{g}_{pro} = \arg\min \mathbf{g}^H \mathbf{R}_o(N)\mathbf{g}, \text{ subject to } \mathbf{g}^H \mathbf{R}_s(N)\mathbf{g} = 1 \quad (4.45)$$

If $\mathbf{R}_s(N)$ is symmetric positive definite, which is generally the case for sufficiently large N and wideband noise, the minimization problem in (4.45) can be converted into:

$$\mathbf{g}_{pro} = \arg\min_{\mathbf{g}} \frac{\mathbf{g}^H \mathbf{R}_o(N)\mathbf{g}}{\mathbf{g}^H \mathbf{R}_s(N)\mathbf{g}} \quad (4.46)$$

Clearly, the desired equalizer is the eigenvector corresponding to the minimal generalized eigenvalue of matrix pair $(\mathbf{R}_o(N), \mathbf{R}_s(N))$. In the following, we shall refer to this equalizer as the downlink C-MOE receiver.

The C-MOE equalizer mimics the ZF solution where existence condition is the well-known "shape length and zero" condition. Loosely speaking, FIR ZF equalization of FIR channels is possible when $M > 1$ [23–27]. We observe, without a rigorous and complete analysis, that if the following mild conditions are satisfied, the proposed equalizer converges to the exact ZF solution, almost always, in noise-free cases.

- All the unused spreading codes are used to construct $\mathbf{C}_o$.

- Equalized channel length satisfies $L + K - 1 < M$.

- N is large enough, typically $N = 10$ suffices.

Asymptotic Performance

We now turn our attention to the asymptotic performance of the C-MOE receiver when $N \to \infty$ and compare it with that of the multiuser MMSE receiver and the ZF equalizer followed by a despreading operation (ZF-DS). We assume both the MMSE decorrelating receiver and the ZF equalizer are constructed with perfect knowledge of the channels.

We use the MSE of the recovered symbol sequence as the performance measure. It is easy to show that the MSE expression for the MMSE decorrelating receiver is:

$$\text{MSE}_{\text{MMSE}} = 1 - \bar{\mathbf{w}}_1^H \mathbf{R}_{\mathbf{yy}}^{-1} \bar{\mathbf{w}}_1$$

where $\bar{w}_1(l) = c_1(l) * h_1(l)$ is the effective signature waveform, and $\mathbf{R}_{\mathbf{yy}} = E[\mathbf{y}(n)\mathbf{y}^H(n)]$ is the covariance matrix of $\mathbf{y}(n)$.

The ZF equalizer, on the other hand, is given by:

$$\mathbf{g}_{\text{zf}} = \mathbf{H}^\dagger \mathbf{e}_{n_0+1}$$

where $\mathbf{H} = [\mathcal{T}(\mathbf{h}_1) \cdots \mathcal{T}(\mathbf{h}_M)]$, $\dagger$ denotes pseudoinverse, and $\mathbf{e}_{n_0+1}$ is the $n_0 + 1$th column of identity matrix $\mathbf{I}_{MK}$. For the ZF-DS receiver,

$$\begin{aligned}
\text{MSE}_{\text{ZF}} &= \frac{1}{M^2} \mathbf{g}_{\text{zf}}^H \mathcal{H}(\mathbf{c}_1^o) \mathbf{R}_{I,n} \mathcal{H}(\mathbf{c}_1^o)^H \mathbf{g}_{\text{zf}} \\
&= \frac{1}{M^2} \mathbf{g}_{\text{zf}}^H \mathcal{H}(\mathbf{c}_1^o) \mathbf{R}_{nn} \mathcal{H}(\mathbf{c}_1^o)^H \mathbf{g}_{\text{zf}}
\end{aligned}$$

where $\mathcal{H}(\mathbf{c}_1^o)$ is a block diagonal matrix with M identical diagonal blocks of Hankel matrices $\mathbf{H}(\mathbf{c}_1^o)$,

$$\mathbf{H}(\mathbf{c}_1^o) = \begin{bmatrix} c(1) & c(2) & \cdots & c(l_t + l_h + M) \\ \vdots & & \ddots & \vdots \\ c(K) & c(K+1) & \cdots & c(l_t + l_h + M + K - 1) \end{bmatrix}$$

$\mathbf{R}_{I,n}$ is the interference-plus-noise covariance matrix and $\mathbf{R}_{nn}$ is noise covariance matrix.

The MSE at the output of the blind C-MOE receiver can be numerically computed as follows. First, we obtain the C-MOE equalizer from (4.46) with $\mathbf{R}_s(N)$ and $\mathbf{R}_o(N)$ substituted by their asymptotic values. Next, we normalized the equalizer such that the desired user's signal energy after equalization and despreading is unity. The MSE is then calculated as:

$$\text{MSE}_{\text{CMOE}} = \mathbf{g}_{\text{pro}}^H \mathcal{H}(\mathbf{c}_1^o) \mathbf{R}_{I,n} \mathcal{H}(\mathbf{c}_1^o)^H \mathbf{g}_{\text{pro}} \quad (4.47)$$

<u>Example 4</u>: For illustration purposes, we compute the MSEs of different receivers for a 10-user CDMA system with spreading gain of 16. All FIR channels span 3 chips, and the equalizer length K is fixed at 4. We plot in Figure 4.9 the output MSEs for three aforementioned linear receivers for a range of SNR values. It is observed that the performance of the C-MOE receiver and the trained ZF receiver are very close. For medium to high SNR, the performance of the C-MOE receiver is almost identical to that of the ZF receiver. The trained MMSE decorrelating receiver serves as a lower bound for all linear receivers. The gap between the MMSE and the C-MOE receiver decreases as SNR increases. This result is consistently observed for a large selection of channels.

4.3.2 Adaptive Implementation

In this section, we derive an adaptive procedure for obtaining $\mathbf{g}_{\text{pro}}$ from (4.45). Some aspects of its convergence are also studied.

To derive the adaptation rules, we first convert (4.45) into an unconstrained minimization problem using the method of Lagrange multipliers.

$$J(N) = \mathbf{g}^H \mathbf{R}_o(N) \mathbf{g} - \lambda(\mathbf{g}^H \mathbf{R}_s(N) \mathbf{g} - 1) \quad (4.48)$$

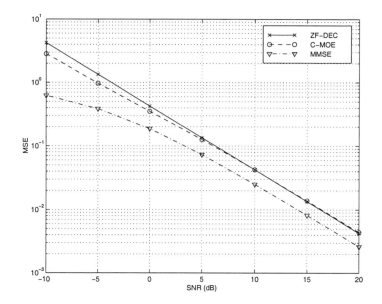

Figure 4.9 Asymptotic performance of three linear receivers.

The gradient of $J(N)$ with respect to $\mathbf{g}$ is given by:

$$\nabla_{\mathbf{g}} J(N) = 2[\mathbf{R}_o(N)\mathbf{g} - \lambda \mathbf{R}_s(N)\mathbf{g}] \quad (4.49)$$

Differentiating $J(N)$ with respect to λ yields constraint $\mathbf{g}^H \mathbf{R}_s(N)\mathbf{g} = 1$. Since we are only interested in the direction of $\mathbf{g}$, $\mathbf{g}$ can be normalized without loss of optimality, $\mathbf{g} = \mathbf{g}/\|\mathbf{g}\|^2$. At the optimum point, the gradient must be a zero vector: $\mathbf{R}_o(N)\mathbf{g} - \lambda \mathbf{R}_s(N)\mathbf{g} = \mathbf{0}$. Premultiplying both sides of the equation by $\mathbf{g}^H$ yields:

$$\lambda = \frac{\mathbf{g}^H \mathbf{R}_o(N)\mathbf{g}}{\mathbf{g}^H \mathbf{R}_s(N)\mathbf{g}} \quad (4.50)$$

To accommodate a possible time varying environment, we adopt the following rules for updating $\mathbf{R}_o(n)$ and $\mathbf{R}_s(n)$,

$$\begin{aligned}
\mathbf{R}_o(n+1) &= (1-\alpha)\mathbf{R}_o(n) \\
&\quad + \alpha \mathbf{Y}^H(n+1)\mathbf{C}_o(n+1)\mathbf{C}_o^H(n+1)\mathbf{Y}(n+1) \\
\mathbf{R}_s(n+1) &= (1-\alpha)\mathbf{R}_s(n) \\
&\quad + \alpha \mathbf{Y}^H(n+1)\mathbf{c}_1^o(n+1)\mathbf{c}_1^o(n+1)^H \mathbf{Y}(n+1)
\end{aligned} \quad (4.51)$$

where $0 < \alpha < 1$ is a smoothing factor. (4.51) is referred to as the exponential windowed time average of $\mathbf{R}_o(n)$ and $\mathbf{R}_s(n)$.

The online implementation of the algorithm involves a recursive update of the output covariance matrices $\mathbf{R}_o(n)$ and $\mathbf{R}_s(n)$, as in (4.51). The adaptive rules are summarized as follows:

1. Initialize parameters.

2. Update $\mathbf{R}_o(n)$, $\mathbf{R}_s(n)$ using (4.51).

3. $\nabla(n) = \mathbf{R}_o(n)\mathbf{g}(n) - \lambda(n)\mathbf{R}_s(n)\mathbf{g}(n)$.

4. $\mathbf{g}(n+1) = \mathbf{g}(n) - \mu\nabla(n)$.

5. $\lambda(n+1) = \dfrac{\mathbf{g}^H(n+1)\mathbf{R}_o(n)\mathbf{g}(n+1)}{\mathbf{g}^H(n+1)\mathbf{R}_s(n)\mathbf{g}(n+1)}$,
   $\mathbf{g}(n+1) = \mathbf{g}(n+1)/\|\mathbf{g}(n+1)\|$.

6. Go back to second step.

In practice, complexity at each iteration can be reduced by using instantaneous estimates:

$$\mathbf{R}_o(n) = \mathbf{Y}^H(n)\mathbf{C}_o(n)\mathbf{C}_o^H(n)\mathbf{Y}(n), \quad \mathbf{R}_s(n) = \mathbf{Y}^H(n)\mathbf{c}_1^o(n)\mathbf{c}_1^o(n)^H\mathbf{Y}(n)$$

in the second step.

Convergence Analysis

The transient behavior of the adaptive C-MOE equalizer is quite involved. Convergence analysis of similar adaptive methods can be found in [7, 28].

Let $\mathbf{R}_o = \lim_{n\to\infty} \mathbf{R}_o(n)$ and $\mathbf{R}_s = \lim_{n\to\infty} \mathbf{R}_s(n)$. Assume:

1. $\mathbf{R}_o$ and $\mathbf{R}_s$ are both symmetric positive definite. This is true with probability one for wideband noise.

2. The smallest generalized eigenvalue of matrix pencil $(\mathbf{R}_o, \mathbf{R}_s)$ has unit multiplicity.

It can be shown that:

1. The algorithm is locally convergent. If **g** is initialized sufficiently close to a generalized eigenvector of $(\mathbf{R}_o, \mathbf{R}_s)$, it will asymptotically converge to it.
2. The undesired eigenvectors are locally unstable while the minimal eigenvector is locally stable. This is because undesired eigenvectors correspond to either saddle points or a maximal point.

Therefore, the adaptive solution asymptotically converges to the minimal generalized eigenvector with probability one when step size μ is chosen properly. It enjoys a global convergence property. In the next subsection, we study how to determine μ in order to ensure convergence.

Step Size

For simplicity, we will study the effect of step size μ on convergence based on the asymptotic covariance matrices $\mathbf{R}_o$ and $\mathbf{R}_s$. We establish an asymptotic upper-bound for μ, which turns out to be useful even for a finite number of data samples.

Proposition 6 *When:*

$$\mu < \frac{2}{\max_i(\lambda_{s,i}(\lambda_i - \lambda_{JK}))}, \quad i = 1, \cdots, MK - 1, \quad (4.52)$$

the adaptive algorithm will converge to the minimal eigenvector with probability one. Here $\lambda_1 \geq \cdots \geq \lambda_{JK}$ is the generalized eigenvalues of matrix pair $(\mathbf{R}_o, \mathbf{R}_s)$, and $\lambda_{s,i}$ is the corresponding eigenvalue of $\mathbf{R}_s$.

Proof: When n is large enough, we can approximate $\mathbf{R}_o(n) \to \mathbf{R}_o$, $\mathbf{R}_s(n) \to \mathbf{R}_o$.

Rewrite the recursion rule as follows,

$$\begin{aligned} \mathbf{g}(n+1) &= \mathbf{g}(n) - \mu[\mathbf{R}_o\mathbf{g}(n) - \lambda(n)\mathbf{R}_s\mathbf{g}(n)] \\ &= [\mathbf{I} - \mu(\mathbf{R}_o - \lambda(n)\mathbf{R}_s)]\mathbf{g}(n) \end{aligned} \quad (4.53)$$

Since both $\mathbf{R}_o$ and $\mathbf{R}_s$ are symmetric positive definite, it is shown in [10] that they can be jointly diagonalized by an orthonormal matrix $\mathbf{V}$ such that:

$$\mathbf{V}^H \mathbf{R}_o \mathbf{V} = \underbrace{\text{diag}(\lambda_{o,1}, \cdots, \lambda_{o,JK})}_{\Lambda_o}$$
$$\mathbf{V}^H \mathbf{R}_s \mathbf{V} = \underbrace{\text{diag}(\lambda_{s,1}, \cdots, \lambda_{s,JK})}_{\Lambda_s} \quad (4.54)$$

and $\lambda_i = \lambda_{o,i}/\lambda_{s,i}$.

Applying the above result to (4.53), we have:

$$\mathbf{g}(n+1) = \mathbf{V}[\mathbf{I} - \mu(\Lambda_o - \lambda(n)\Lambda_s)]\mathbf{V}^H \mathbf{g}(n) \quad (4.55)$$

$\mathbf{g}(n)$ can be expressed as a linear combination of column vector $\mathbf{V}$,

$$\mathbf{g}(n) = \sum_{i=1}^{MK} \mathbf{v}_i t_i(n) \quad (4.56)$$

Combining (4.55) and (4.56), we obtain the updated equations for the weighting coefficients:

$$\begin{aligned} t_i(n+1) &= [1 - \mu(\lambda_{o,i} - \lambda(n)\lambda_{s,i})]t_i(n) \\ &= [1 - \mu\lambda_{s,i}(\lambda_i - \lambda(n))]t_i(n) \\ &= \beta_i(n)w_i(n), \quad i = 1, \cdots, MK \end{aligned} \quad (4.57)$$

$\lambda_i(n)$ satisfies $\lambda_1 \geq \lambda(n) \geq \lambda_{JK}$. For those $\lambda_m > \lambda(n)$, the weight gain $\beta_m(n) \leq 1$. While for those $\lambda_n < \lambda(n)$, weight gain $\beta_n(n) \geq 1$. Therefore, if we can make sure $\beta_i(n) > -1$, for $i = 1, \cdots, MK$, then after the current iteration, the components in $\mathbf{g}(n)$ corresponding to eigenvectors with eigenvalues that are larger than $\lambda(n)$ will deflate, while components corresponding to smaller eigenvalues will inflate. Therefore, $\lambda(n+1) \leq \lambda(n)$. This will drive $\lambda(n)$ toward the smallest eigenvalue λ_{MK}. Due to this deflation-inflation process, components corresponding to eigenvectors associated with eigenvalues that are greater than λ_{MK} will eventually die out with $-1 < \beta_i(\infty) < 1$. Only the eigenvector corresponding to λ_{MK} sustains with $\beta_{MK}(\infty) = 1$.

The upper-bound for μ can be obtained from the requirement that $\beta_i(n) \geq -1$. Further relaxing:

$$\beta_i(n) \geq 1 - \mu\lambda_{s,i}(\lambda_i - \lambda_{MK}) \geq 1 \qquad (4.58)$$

we reach the upper-bound given in (4.52).

The convergence speed of the adaptive implementation depends primarily on the value of $\lambda_{s,MK}$, and the difference between λ_{MK-1} to λ_{MK}. The larger the value of $\lambda_{s,MK}$ corresponds to faster convergence speed. The bigger the difference between the two smallest eigenvalues of $(\mathbf{R}_o, \mathbf{R}_s)$, the quicker the algorithm converges.

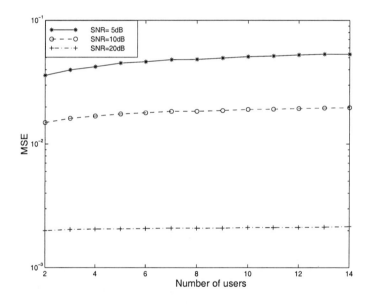

Figure 4.10 MSEs versus number of of users.

4.3.3 Examples

We now provide some numerical results to illustrate the behavior of the C-MOE receiver under different circumstances. The performance of the blind C-MOE receiver will be compared with the ideal ZF receiver calculated with perfect knowledge of the channel coefficients. In all simulations, a pseudorandom mask code of $\{-1,1\}$ is employed to overlay the entire data block.

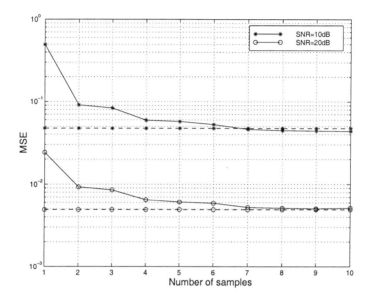

Figure 4.11 MSEs versus number of samples.

Example 5: First, we study the effect of system load on the performance of the proposed receiver. We simulate a CDMA system with spreading gain of 16. Two FIR channels of length 3 are used, and the input SNR is set at 10 dB. The length of the equalizer for both channels is 3. Fifty symbols are used to compute the blind channel equalizer. We measure the MSEs of receiver output for a range of system load. Figure 4.10 shows the results. The three curves correspond to three SNR levels. Overall, the proposed equalizer is robust against increases in the number of total users. The curve flats out with increased SNR values. Note that the ZF receiver is immune to changes in system load.

Example 6: The data efficiency of the proposed equalizer is studied in the next example. We choose the CDMA system used in example 5 except that the number of users P is set to be 10. We plot in Figure 4.11 the MSE of symbol estimates after the despreading operation

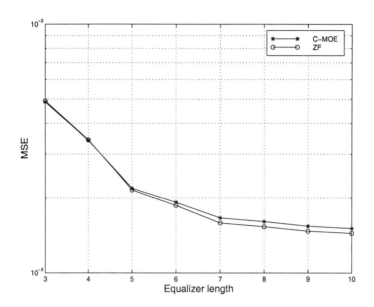

Figure 4.12 Performance over length of equalizer.

over different N for two SNR values. The two dashed lines represent the performance of the ideal ZF receiver. The results show that the C-MOE solution is very data efficient. It converges to the ZF receiver for only 7 symbol vectors. In most of the simulations conducted, about 10 data vectors (symbols) are sufficient to compute a reliable equalizer.

Example 7: Next, we study the effect of equalizer length on its performance. We vary the length of the equalizer while keeping delay fixed at $n_0 = L = 3$. Figure 4.12 shows that the performance of the ZF and the C-MOE receivers improves with increasing length of equalizer. For large K, the curves flat out. Therefore, it is not necessary to choose K to be very large.

In Figure 4.13, we give an example of the equalized channel impulse response. We use 2 FIR channels of length 5 and 2 equalizers of length 6. At 10 dB SNR, 40 received data vectors are used to obtain equalizer $g_1(n)$ and $g_2(n)$. In Figure 4.13.(a), we plot the real part and imaginary part of the channel response, $h_1(n)$ and $h_2(n)$. The equalized channel response $f(n)$ is illustrated in Figure 4.13.(b).

CDMA With Long Codes: Space-Time Processing

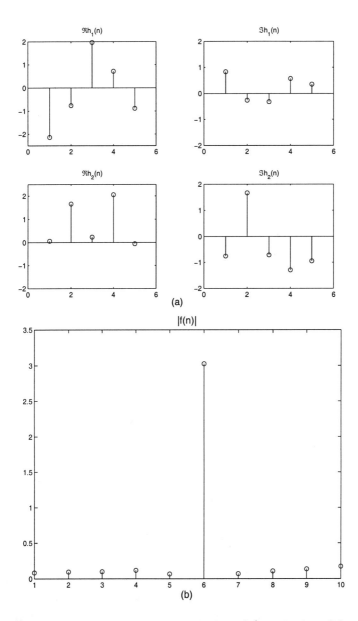

Figure 4.13 Channel impulse response before (a) and after (b) equalization.

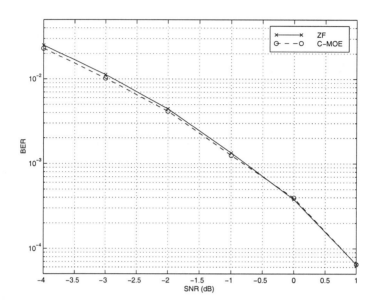

Figure 4.14 BER versus SNR.

Example 8: In this example, we measure the bit error rate (BER) of the ZF receiver and the proposed blind receiver under a fixed channel. The configuration is identical to that of Example 7. DQPSK signals are used. Differential decoding can resolve the scalar ambiguity of blind equalizer from solving (4.46). We plot the BER of the two receivers in Figure 4.14. The two curves are almost identical. This shows that the BER performance of the blind C-MOE receivers is close to that of the ZF receiver that has exact knowledge of the channel response.

Example 9: In this experiment, the adaptive version of the proposed receiver is investigated. We measure the performance of the adaptive algorithm using instantaneous estimates of $\mathbf{R}_o(n)$ and $\mathbf{R}_s(n)$ and sample averages of $\mathbf{R}_o(n)$ and $\mathbf{R}_s(n)$, respectively. We track the normalized MSEs of two implementations with data blocks of size 200. We then compare them with the MSE obtained by the batch method. The step size μ is set to 2×10^{-4}. SNR is 10 dB. In Figure 4.15, we can see that the MSE of adaptive solution approaches quickly to the straight line at the bottom, which denotes the MSE of batch-mode im-

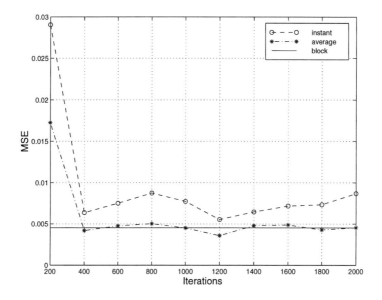

Figure 4.15 Performance of adaptive receiver.

plementation. Numerical result shows that the adaptive algorithm can converge to the desired equalizer with only a small amount of excess MSE. Compared with the more complex "averaging" implementation, the "instant" method converges slightly slower and has larger excessive steady-state errors.

4.4 Conclusion

In this chapter, we focused on the problem of blind detection of CDMA signals that are modulated with long spreading codes. We first discussed the problem in the context of 2D RAKE receivers. The situation arises in uplink CDMA where the base station employs an array of antennas. A class of 2D RAKE reception schemes have been presented. Although the same approaches can be used in downlink, a more appealing strategy is to equalize the "common" channel and restore the orthogonality before extracting the desired signal via despreading.

References

[1] R. Price and P. E. Green, "A communication technique for multipath channels," *Proc. IRE*, **46**:555–570, March 1958.

[2] A. F. Naguid and A. Paulraj, "A base-station antenna array receiver for cellular DS/CDMA with M-ary orthogonal modulation," In *Proc. 28nd Asilomar Conf. on Signals, Systems, and Computers*, pages 858–862, Pacific Grove, CA, November 1994.

[3] E. G. Ström, S. Parkvall, S. L. Miller, and B. E. Ottersten, "Propagation delay estimation in asynchronous direct-sequence code-division multiple-access systems," *IEEE Trans. on Communications*, **44**(1):84–93, January 1996.

[4] M. D. Zoltowski and J. Ramos, "Blind adaptive beamforming for CDMA-based PCS/Cellular," In *Proc. 30nd Asilomar Conf. on Signals, Systems, and Computers*, Pacific Grove, CA, November 1995.

[5] A. Klein, "Data detection algorithms specially designed for the downlink of CDMA mobile radio systems," In *Proc. VTC97*, pages 203–207, Phoenix, AZ, May 1997.

[6] I. Ghauri and D. T. M. Slock, "Linear receiver for the DS-CDMA downlink exploiting rrthogonality of spreading sequences," In *Proc. 32th Asilomar Conference on Signals, Systems, and Computers*, Pacific Grove, CA, November 1998.

[7] G. Mathew and V. U. Reddy, "A Quasi-Newton adaptive algorithm for generalized symmetric eigenvalue problem," *IEEE Trans. on Signal Processing*, **SP-44**(10):2413–2422, October 1996.

[8] H. Chen, T. K. Sarkar, S. A. Dianat, and J. D. Brule, "Adaptive spectral estimation by the conjugate gradient method," *IEEE Trans. on ASSP*, **ASSP-34**(2):272–284, April 1986.

[9] P. Strobach, "Fast orthogonal iteration adaptive algorithms for the generalized symmetric eigenproblem," *IEEE Trans on Signal Processing*, **SP-46**(12):3345–3359, December 1998.

[10] G. Golub and C. Van Loan, *Matrix computations*, third edition, Baltimore, MD: Johns Hopkins University Press, 1996.

[11] H. Liu, G. Xu, L. Tong, and T. Kailath, "Recent developments in

blind channel equalization: From cyclostationarity to subspaces," *Signal Processing*, vol. 50, pages 83–99, June 1996

[12] L. Tong, G. Xu, and T. Kailath, "Blind identification and equalization using spectral measures, Part II: A time domain approach," In W.A. Gardner, editor, *Cyclostationarity in Communications and Signal Processing*. IEEE Press, 1993.

[13] E. Moulines, P. Duhamel, J. Cardoso, and S. Mayrargue, "Subspace methods for the blind identification of multichannel FIR filters," *IEEE Trans. on Signal Processing*, **SP-43**(2):516–525, February 1995.

[14] G. B. Giannakis and S. D. Halford, "Blind fractionally-spaced equalization of noisy FIR channels: Adaptive and optimal solutions," *IEEE Trans. on Signal Processing*, pages 2277–2292, September 1997.

[15] D. T. M. Slock and C. B. Papadias, "Further results on blind identification and equalization of multiple FIR channels," In *Proc. IEEE ICASSP*, pages 1964–1967, May 1995.

[16] G. Xu, H. Liu, L. Tong, and T. Kailath, "A least-squares approach to blind channel identification," *IEEE Trans. on Signal Processing*, **SP-43**(12):2982–2993, December 1995.

[17] Y. Hua, "Fast maximum likelihood for blind identification of multiple FIR channels," In *Proc. 28th Asilomar Conf. on Signals, Systems, and Computers*, Pacific Grove, CA, November 1994.

[18] B. Suard, A.F. Naguib, G. Xu, and A. Paulraj, "Performance of CDMA Mobile Communication Systems Using Antenna Arrays," In *Proc. ICASSP Conf.*, Minneapolis, MN, April 1993.

[19] Tan F. Wong, Tat M. Lok, James S. Lehnert, and Michael D. Zoltowski, "A unified linear receiver for direct-sequence spread-spectrum multiple-access systems with antenna arrays and blind adaptation," submitted to *IEEE Trans. on Information Theory*, February 1996.

[20] Tan F. Wong, Tat M. Lok, James S. Lehnert, and Michael D. Zoltowski, "Spread-spectrum signaling techniques with antenna arrays and blind adaptation," accepted for presentation in special session at *Milcom '96,* October 1996.

[21] J. Yang and A. Swindlehurst, "DF-directed multipath equaliza-

tion," In *Proc. 28nd Asilomar Conf. on Signals, Systems, and Computers*, pages 1418–1422, Pacific Grove, CA, November 1994.

[22] J. G. Proakis, *Digital Communications*, third edition, New York: McGraw-Hill, 1995.

[23] D. T. M. Slock, "Blind fractionally-spaced equalization, perfect-reconstruction filter banks and multichannel linear prediction," In *Proc. IEEE ICASSP*, pages IV585–IV588, Adelaide, Australia, April 1994.

[24] L. Tong, G. Xu, and T. Kailath, "Blind identification and equalization based on second-order statistics: A time domain approach," *IEEE Trans. on Information Theory*, **IT-40**(2):340–349, March 1994.

[25] G. B. Giannakis and S. D. Halford, "Blind fractionally-spaced equalization of noisy FIR channels: Direct and adaptive solutions," *IEEE Trans. on Signal Processing*, **SP-45**(9):2277–2292, September 1997.

[26] E. Moulines, P. Duhamel, J. Cardoso, and S. Mayrargue, "Subspace methods for the blind identification of multichannel FIR filters," *IEEE Trans. on Signal Processing*, **SP-43**(2):516–525, February 1995.

[27] Y. Li and K. J. Ray Liu, "Blind adaptive equalization and diversity combining," In *Proc. ICASSP*, Munich, Germany, April 1997.

[28] Z. Xu and M. K. Tsatsanis, "Adaptive minimum variance methods for direct blind multichannel equalization," In *Proc. IEEE ICASSP*, pages 2105–2108, Seattle, WA, May 1998.

Chapter 5

Multicarrier CDMA

In Chapters 2–5 our attempts at dealing with the multipath channel effect have focused on the receiver end. Our point of view is that broadband DS-CDMA signals, when transmitted through frequency-selective fading channels, are inevitably distorted and thus can only be recovered through receiver side processing.

Instead of relying on sophisticated reception algorithms, the technique of "multicarrier transmission" copes with hostile radio channels with low-speed, parallel operations. By transmitting high-speed data through low-rate streams, the symbol duration in each substream increases, leading to higher immunity against multipath dispersion. The advantages and success of multicarrier modulation (MCM) and CDMA techniques motivated many researchers to investigate the suitability of combining MCM with CDMA for wideband multiple-access communications [1–9]. This combination, known as MC-CDMA, allows one to benefit from the advantages of both schemes.

Several schemes have been proposed to date. An overview of the three most popular proposals, multitone CDMA (MT-CDMA) [10], multicarrier direct-sequence CDMA (MC-DS-CDMA) [11, 12, 13], and multicarrier CDMA (MC-CDMA) [14, 15], is given in [16]. In all three schemes, users are allowed to transmit on many available subchannels, thus obtaining the maximum benefit from multicarrier transmission. Each user is assigned a CDMA code, which is used to differentiate between signals belonging to different users at the receiver.

MC-CDMA could be particularly useful in the uplink of a mobile system, since it retains many of the benefits of CDMA, such as soft handoff and high-spectral efficiency, while relaxing the stringent (sometimes impossible) synchronization requirements in broadband systems. However, multifaceted challenges remain before MC-CDMA can achieve its full potential. One of the open areas is in receiver design. Another issue is the sensitivity of MC-CDMA toward random carrier offsets. While it is widely known that the effect of frequency dispersion on MCM is similar to that of time dispersion on single-carrier transmission, there have been surprisingly few contributions on this topic.

In this chapter, we will first study the modulation and detection strategies of MC-CDMA based on a common framework that encompasses several existing MC-CDMA schemes. The sensitivity of MC-CDMA in the presence of carrier offsets will then be investigated. In particular, we will derive the analytical expressions for the degradation in SINR caused by interchannel interference (ICI) and multiple-access interference (MAI).

5.1 MC-CDMA Overview

5.1.1 Multicarrier and OFDM

To understand MC-CDMA, we must first review the technique of multicarrier modulation for single-user high-speed communications. MCM is the principle of transmitting high-rate data by dividing incoming data into many parallel bit streams, each of which has a much lower bit rate. The simplest MCM comes in the form of the standard frequency division multiplexing (FDM), where incoming bit sequence is serial-to-parallel converted and transmitted through low-rate, nonoverlapping subchannels; see Figure 5.1. The narrowband nature of signals modulated on each subcarrier provides a high immunity against multipath dispersion and narrowband interference.

Despite its simple concept, MCM did not draw significant attention in high-speed communication until recent years. The reasons are two-fold. First, conventional FDM requires steep bandpass filters that

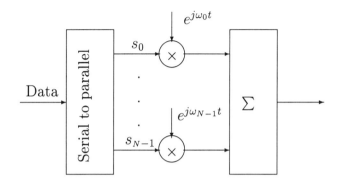

Figure 5.1 MCM modulation.

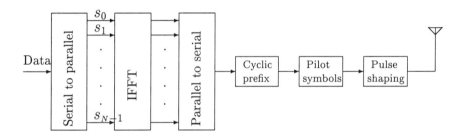

Figure 5.2 OFDM modulator.

incur considerable cost when the number of subchannels is large. Second, the guardband in nonoverlapping FDM usually results in a certain loss in spectrum efficiency that is undesirable in most resource-critical applications.

The first problem can be dealt with by using faster DSPs and filterbank techniques that exploit the parallelism of structured filters. A more efficient solution to both problems is the so-called orthogonal frequency division multiplexing (OFDM), which can be realized by using the discrete Fourier transform. Figure 5.2 illustrates the basic modules of an OFDM modulator. In OFDM, high-speed data is modulated in blocks using the IFFT. The output signals of the IFFT have the overlapping sinc spectra centered at a bank of dense subcarriers as shown in Figure 5.3. To combat the multipath channel effect, each block of signals is appended with a "cyclic prefix" of length longer than the maximum delay spread of the channel. To explain this idea of guard period, let us denote N as the IFFT block size, N_g the length of the cyclic prefix, $W = 2\pi/N$, and $s_0 \cdots s_{N-1}$ the block of signals to be modulated. The $N + N_g$ samples (after pending the cyclic prefix) of OFDM modulated signals can be mathematically expressed as:

$$\begin{bmatrix} 1 & W^{N-N_g} & \cdots & \cdot & W^{(N-N_g)(N-1)} \\ \vdots & \vdots & \vdots & & \vdots \\ 1 & W^{N-1} & \vdots & \vdots & W^{(N-1)^2} \\ - & - & - & - & - \\ 1 & 1 & \cdots & 1 & 1 \\ 1 & W^1 & \vdots & \vdots & W^{N-1} \\ 1 & W^2 & \cdots & \vdots & W^{2(N-1)} \\ \vdots & \vdots & \vdots & \vdots & \vdots \\ 1 & W^{N-1} & \cdots & \cdot & W^{(N-1)^2} \end{bmatrix} \begin{bmatrix} s_0 \\ s_1 \\ \vdots \\ s_{N-1} \end{bmatrix}$$

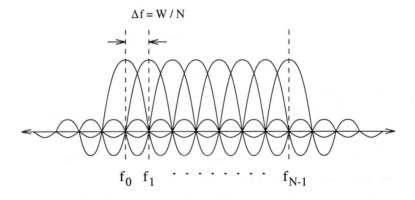

Figure 5.3 OFDM transmit power spectra.

At the receiver end, the output of the channel is the convolution of the transmitted sequence with the channel impulse response. Let us denote the frequency response of the multipath channel (with maximum length N_g) as:

$$H(i) = \sum_{n=0}^{N_g-1} h(n) W^{-i\,n}$$

The removal of the prefix at the receiver will make this operation look like circular convolution [9]. In particular, resulting N samples are of form:

$$\mathbf{x} = \begin{bmatrix} x_0 \\ \vdots \\ x_{N-1} \end{bmatrix} = \mathbf{W}^H \begin{bmatrix} H(0)s_0 \\ \vdots \\ H(N-1)s_{N-1} \end{bmatrix} \quad (5.1)$$

where $\mathbf{W}^H$ is an N-point IFFT matrix. Converting the received signal back to the frequency domain yields:

$$\mathbf{z} = \begin{bmatrix} z_0 \\ \vdots \\ z_{N-1} \end{bmatrix} = \mathbf{W}\mathbf{x} = \begin{bmatrix} H(0)s_0 \\ \vdots \\ H(N-1)s_{N-1} \end{bmatrix}$$

It is seen that the effect of the channel is a mere "scaling" on each subchannel. Since the scalar ambiguity can be removed with differential decoding, it can be argued that MCM/OFDM is "immune" to the time-dispersion effect and thus has an advantage over single-carrier modulation with a linear equalizer.

It should be pointed out that it is often impractical for OFDM to obviate equalization/receiver-side processing in real applications. The simple scheme depicted in Figure 5.2 has little practical value since:

- Deep fading on some subchannels may render the information bits on these subchannels unusable;

- Frequency dispersion due to carrier offset can cause interchannel interference among the subcarrier.

In order for OFDM to be viable in practice, some forms of diversity/equalization techniques are needed. Commonly used techniques

126 *Signal Processing Applications in CDMA Communications*

include coding cross-subcarriers, power loading, frequency domain spreading, and receiver-side frequency domain equalization.

5.1.2 MC-CDMA

MCM/OFDM can be combined with CDMA in several ways to serve multiple-access communications. Such a combination has the benefits of both MCM and CDMA. Here, we will first review a simple scheme proposed in [14, 15]. For historic reasons, we shall refer to it as MC-CDMA from now on.[1]

Figure 5.4 illustrates the OFDM version of a MC-CDMA modulator for user k. Each symbol to be transmitted is spread over M subchannels using a given spreading code in the frequency domain. The number of subcarriers M is usually less than or equal to the IFFT size N when extra spacing between subchannels is needed. The use of OFDM necessitates the use of ISI-preventing cyclic prefix but allows optimum use of the available bandwidth. Alternatively, one can adopt nonoverlapping FDM without the guard time, provided that the symbol duration is significantly larger than the time dispersion of the channel.

The MC-CDMA scheme is actually a generalized version of the frequency-hopping spread spectrum discussed in Chapter 1 – instead of each user using only one subchannel, symbols in MC-CDMA are modulated on many subcarriers to introduce the frequency diversity. For this reason, MC-CDMA is robust against deep selective fading as regular DS-CDMA. The symbol rate of the modulator in Figure 5.4 defines the "base rate" of MC-CDMA transmission. For higher-speed applications, one can simply assign more spreading codes to the user to achieve a rate that is a multiple of the base rate.

Signals from different users are symbol level synchronized in MC-CDMA. Following the data model of OFDM in (5.1), it is not difficult to arrive at the following expression of the n block of MC-CDMA

---

[1]The term *MC-CDMA* has been used in literature to refer to both the general concept of MCM-based CDMA and the specific scheme proposed in [14, 15].

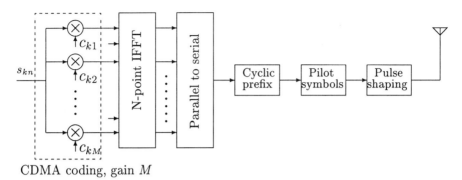

Figure 5.4 Joint model of transmitter and channel for user k.

signals,

$$\mathbf{x}(n) = \begin{bmatrix} x_0(n) \\ \vdots \\ x_{N-1}(n) \end{bmatrix} = \sum_{k=1}^{P} \mathbf{W}^H \mathbf{a}_k s_{k,n} + \mathbf{n}(n) \quad (5.2)$$

where:

$$\mathbf{a}_k = [a_{k,1} a_{k,2} \cdots a_{k,M}]^T, \quad a_{k,m} = c_{k,m} H_k(m)$$

is the "effective spreading vector" (composite effect of the spreading codes and the channel) of the kth user, and $\mathbf{n}(n)$ is the noise vector. The spreading matrix is defined as:

$$\mathbf{A} = [\mathbf{a}_1 \ \mathbf{a}_2 \cdots \mathbf{a}_K]$$

$\mathbf{W}^H$ in (5.2) is a modified modulation matrix of form:

$$\mathbf{W}^H = \begin{bmatrix} 1 & 1 & \cdots & 1 \\ 1 & e^{j2\pi\beta/N} & \cdots & e^{j2\pi(M-1)\beta/N} \\ \vdots & \vdots & \ddots & \vdots \\ 1 & e^{j2\pi(N-1)\beta/N} & \cdots & e^{j2\pi(N-1)(M-1)\beta/N} \end{bmatrix} \quad (5.3)$$

The form of $\mathbf{W}^H$ depends on the number of subcarriers that are used and the spacing between subcarriers. In Figure 5.5, we illustrate the frequency-domain spectra of two multicarrier systems. The first spectrum shows a system that is "tightly packed" in frequency, as in

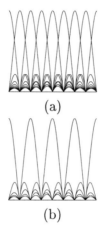

Figure 5.5 Multicarrier frequency spectra. (a) 8 subcarriers, $\beta = 1$; (b) 4 subcarriers, $\beta = 2$.

a standard OFDM system. The second system has subcarriers that are spaced further apart. If the inter-subcarrier spacing $2\pi/M$ is an integer multiple of the subcarrier bandwidth $2\pi/N$, then the spectrum shown in Figure 5.5 (b) can be attained using a transmitter for N-ary transmission with zeros assigned to the unused subcarriers. For a system with M carrier frequencies and subcarrier bandwidth $2\pi/N$, we define the ratio $\beta = N/M$. Thus, $\mathbf{W}^H$ is constructed with an N-ary IDFT matrix with the columns corresponding to unused subcarriers removed.

The capacity of MC-CDMA is limited by the multiple-access interference, as in DS-CDMA, and carrier frequency dispersion-induced ICI. Performance and robustness to frequency offset can then be gained at the price of an increase in computational complexity and bandwidth efficiency. To achieve high performance, channel-dependent multiuser detection is obviously needed. However relative to time-domain DS-CDMA, MC-CDMA does have the following distinctive advantages:

- Synchronization: Block synchronization can be achieved and maintained in MC-CDMA due to the long chip/symbol duration. Such synchronization is instrumental to multiple-user detection.

- Loading: With information being transmitted in parallel narrowband streams, it is convenient to employ adaptive loading techniques to distribute transmission power efficiently based on the subchannel SINR to achieve optimum efficient.

In reception, the received block of signals is first converted back into M parallel streams using the FFT. Define:

$$\mathbf{z}(n) = \mathbf{W}\mathbf{x}(n) = \sum_{k=1}^{P} \mathbf{a}_k s_{k,n} + \mathbf{W}\mathbf{n}(n) \stackrel{\text{def}}{=} \sum_{k=1}^{P} \mathbf{a}_k s_{k,n} + \mathbf{v}(n) \quad (5.4)$$

In order to recover the desired signal from user j, $s_j(n)$, the receiver will combine information from all subchannels using an $M \times 1$ weight vector $\mathbf{b}_j = [b_{j,1} \cdots b_{jM}]^T$:

$$\hat{s}_{j,n} = \mathbf{b}_j^H \mathbf{z}(n) = \mathbf{b}_j^H \left(\sum_{k=1}^{P} \mathbf{a}_k s_{k,n} + \mathbf{v}(n) \right)$$

The choice of weight vector depends on the type of detector that is implemented. The most commonly used detectors are the minimum mean-squared error (MMSE) multiuser detector and the coherent combining detector (CCR). Let $\mathbf{s}_n = [s_{1,n} \cdots s_{P,n}]^T$ and $\hat{\mathbf{s}}_n = [\hat{s}_{1,n} \cdots \hat{s}_{P,n}]^T$ denote the signal vector and its estimate, respectively.

- *The linear MMSE receiver:* For the MMSE receiver, the matrix $\mathbf{B} = [\mathbf{b}_1 \cdots \mathbf{b}_K]$ is chosen so as to minimize $||\hat{\mathbf{s}}_n - \mathbf{s}_n||^2$. Thus $\mathbf{B}$ satisfies the orthogonality principle:

$$\mathcal{E}\left((\hat{\mathbf{s}}_n - \mathbf{s}_n)(\mathbf{A}\mathbf{s}_n + \mathbf{v}_n)^H\right) = 0$$

For white additive noise with variance σ_n^2 and i.i.d. signals with power σ_x^2, it can be shown that $\mathbf{B}$:

$$\mathbf{B}^H = \mathbf{A}^H \left(\mathbf{A}\mathbf{A}^H + (\sigma_n^2/\sigma_x^2)\mathbf{I} \right)^{-1} \quad (5.5)$$

- *The coherent combining receiver:* Sometimes the information for constructing the MMSE receiver may not be available or the MMSE detection may be too expensive to implement. In these cases, the detection matrix for the CCR simply weights the output signal by the channel coefficients for the user being received. While mostly suboptimum, the CCR minimizes the mean-square error $|\hat{s}_{j,n} - s_{j,n}|^2$ when both the inference and noise is white. The detector matrix for the CCR receiver is $\mathbf{B}^H = \mathbf{A}^H$.

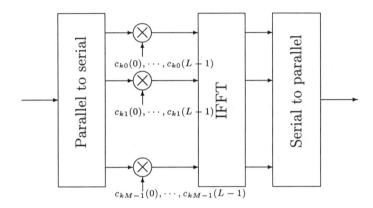

Figure 5.6 MC direct-sequence CDMA modulator.

5.1.3 MC Direct-Sequence CDMA

Multicarrier DS-CDMA [11, 12] is an enhanced version of MC-CDMA where additional time-domain spreading is injected at the transmitter. Figure 5.6 shows the MC-DS-CDMA transmitter for user k. Instead of spreading the data in frequency-domain directly, each symbol is first spread by a factor of L as in DS-CDMA. This extra spreading facilitates joint exploitation of the frequency and time diversities in a multiple-access environment. To ensure frequency nonselective fading on each subchannel, the time spreading factor L has to be appropriately chosen in practice.

In the original MC-DS-CDMA proposal, the same spreading codes are used on all subchannels for each user (i.e., $c_{i1}(l) = \cdots = c_{iM}(l)$). The framework here allows extra degrees of freedom by using different spread codes on different subchannels. It should become clear later that such adjustment can significantly simplify the demodulation operations.

MC-DS-CDMA can be viewed as "a collection of synchronous narrowband DS-CDMA" signals. Indeed, within each subchannel the received signal is exactly the same as that in a synchronous narrowband CDMA system. The difference, which is critical for narrowband fading/interference resistance, is that signals in MC-DS-CDMA are spread across all subchannels.

The MC-DS-CDMA framework unifies several multicarrier and CDMA proposals to date. For example:

- By using uniform spreading code (i.e., $c_{i1}(l) = \cdots = c_{iM}(l)$), the current system becomes the multicarrier direct-sequence CDMA proposed in [3].

- By setting $L = 1$ (i.e., no spreading in time domain), it reduces to the MC-CDMA scheme.

- By letting all except one element in $\{c_{im}\}_{m=1}^{M}$ to zero, this framework also includes conventional narrowband CDMA systems, where each user transmits through only one subchannel.

In MC-DS-CDMA, the effective spreading gain of the system is ML (due to both frequency and time spreading), allowing it to handle a heavily-loaded system with $P > L$ and M. Accounting for the L spreading chips within each symbol period, we can modify the data model of received signals for MC-CDMA (5.4) to obtain its counterpart for MC-DS-CDMA. The L chip samples of received signal in MC-DS-CDMA on the mth subchannel are given by:

$$\mathbf{y}_m(n) = \begin{bmatrix} y_{m0}(n) \\ y_{m1}(n) \\ \vdots \\ y_{mL-1}(n) \end{bmatrix}$$

$$= \sum_{k=1}^{P} H_k(m) \begin{bmatrix} c_{km}(0) \\ c_{km}(1) \\ \vdots \\ c_{km}(L-1) \end{bmatrix} s_k(n) + \begin{bmatrix} v_m(0) \\ v_m(1) \\ \vdots \\ v_m(L-1) \end{bmatrix}$$

$$\stackrel{\text{def}}{=} \sum_{k=1}^{P} H_k(m) \mathbf{c}_{km} s_k(n) + \mathbf{v}_m(n) \qquad (5.6)$$

Stacking sample vectors from all subcarriers, we arrive at a super vector of received data:

$$\mathbf{y}(k) \stackrel{\text{def}}{=} \begin{bmatrix} \mathbf{y}_1(k) \\ \vdots \\ \mathbf{y}_M(k) \end{bmatrix} = \sum_{i=1}^{P} \begin{bmatrix} H_i(1)\mathbf{c}_{i1} \\ \vdots \\ H_i(M)\mathbf{c}_{iM} \end{bmatrix} s_i(k) + \mathbf{v}(k) \quad (5.7)$$

Similarly we can define the ML effective spreading vector for MC-DS-CDMA as:

$$\mathbf{a}_k = [H_k(0)\mathbf{c}_{k0}^T \cdots H_k(M-1)\mathbf{c}_{kM-1}^T]^T \quad (5.8)$$

The exact formulation used in MC-CDMA follows,

$$\mathbf{y}(n) = \sum_{k=1}^{P} \mathbf{a}_k s_k(n) + \mathbf{v}(n) \stackrel{\text{def}}{=} \mathbf{A}\mathbf{s}(k) + \mathbf{v}(k) \quad (5.9)$$

- *Despreading-combining receiver*: To recover the jth signal from $\{\mathbf{y}_m(n)\}_{m=0}^{M-1}$, it is suggested in [3] to perform demodulation in two steps, namely, despreading and combining. In particular, $\{\mathbf{y}_m(k)\}$ are first despread with the spreading code $\{\mathbf{c}_{jm}\}_{m=0}^{M-1}$ to obtain M symbol sequences:

$$z_{jm}(n) = \sum_{l=1}^{L} c_{jm}(l) y_{ml}(n) = \mathbf{c}_{jm}^H \mathbf{y}_m(n), \quad m = 0, \cdots, M-1 \quad (5.10)$$

The despreading operation essentially reduces the MC-DS-CDMA to MC-CDMA. If we let $\mathbf{z}_j(n) = [z_{j0}(k) \cdots z_{jM-1}(k)]$, the MMSE receiver of MC-CDMA readily applies to $\mathbf{z}_j(n)$:

$$\arg_{\mathbf{b}_{zi}} \min \mathcal{E} |\underbrace{\mathbf{b}_{zj}\mathbf{z}_i(n)}_{\hat{s}_j(n)} - s_j(k)|^2 \quad (5.11)$$

Rewriting $\mathbf{z}_j = \mathbf{C}_j^H \mathbf{y}$ by denoting:

$$\mathbf{C}_j = \begin{bmatrix} \mathbf{c}_{j0} & & \\ & \ddots & \\ & & \mathbf{c}_{jM-1} \end{bmatrix}_{ML \times M} \quad (5.12)$$

it can be shown that the optimum combining vector $\mathbf{b}_{zj}$ is given by the solution of:

$$\left(\mathbf{C}_j^H \mathbf{A} \mathbf{A}^H \mathbf{C}_j + \mathbf{C}_j^H \mathbf{R}_v \mathbf{C}_j\right) \mathbf{b}_{zj} = \mathbf{C}_j^H \mathbf{a}_j \quad (5.13)$$

- *Minimum mean-squared error receiver:*

  On the other hand, the true MMSE receiver for MC-DS-CDMA operates directly on the chip-rate sample signal $\mathbf{y}(k)$ of dimension $LM \times 1$. For simultaneous optimization for all P users, the selection of weight vectors can be compactly expressed as:

  $$\arg_{\mathbf{B}_y} \min \mathcal{E} \| \underbrace{\mathbf{B}_y \mathbf{y}(k)}_{\hat{\mathbf{s}}(k)} - \mathbf{s}(k) \|_F^2, \quad \mathbf{B}_y = [\mathbf{b}_{y1} \cdots \mathbf{b}_{yP}]_{ML \times P} \quad (5.14)$$

  Similarly to (5.13), it can be shown that the $ML \times 1$ weight vector $\mathbf{b}_{yj}$ is given by the solution of:

  $$\left(\mathbf{A} \mathbf{A}^H + \mathbf{R}_v\right) \mathbf{b}_{yj} = \mathbf{a}_j \quad (5.15)$$

Comparing MC-DS-CDMA with MC-CDMA and DS-CDMA, it is obvious that MC-DS-CDMA serves as a compromise between the frequency- and time-domain spreading. Whether MC-DS-CDMA the advantages of both MC-CDMA and DS-CDMA depends critically on the design of the system parameters, such as the frequency- and time-spreading factors. When the bandwidth of the subchannel is fixed, MC-DS-CDMA does offer stronger fading resistance than regular narrowband DS-CDMA and higher degree flexibility than MC-CDMA. The tradeoff, as expected, is a higher computational cost. The DC receiver of MC-DS-CDMA has the same complexity as the MC-CDMA MMSE receiver, since the despreading front-end can be realized with dedicated circuitry. However the MC-DS-CDMA MMSE receiver has significant higher complexity ($O(M^3 L^3)$ and $O(M^2 L^2)$ for batch and adaptive processing, respectively). Unless the cost of MMSE can be reduced, the advantage of MC-DS-CDMA over MC-CDMA cannot be fully realized in practice.

An interesting question to ask here is whether it is possible to reduce the complexity of the MC-DS-CDMA MMSE receiver without sacrificing the performance. Later, we will show that such is indeed possible by exploiting the intrinsic structure of MC-DS-CDMA.

5.1.4 Multitone-CDMA

Another representative proposal in combining MCM with CDMA is the multitone-CDMA [10] where MCM modulation is first performed and the resulting symbols are then spread as in Figure 5.7. The scheme benefits from the resistance of MCM and multiple-access capacity of CDMA, and it is possible to benefit from diversity reception like selection diversity or RAKE reception.

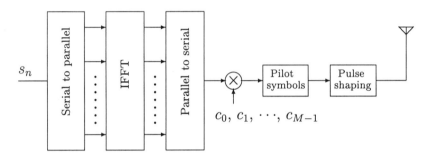

Figure 5.7 Multitone-CDMA modulator.

The time-spreading scrambles the output of MCM and makes the resulting spectrum random as illustrated in Figure 5.8. Given a constant bandwidth, the chip duration has to be constant, as does the ratio between the number of tones and the number of chips per symbol. Since MT-CDMA is closer to regular DS-CDMA, we will focus on MC-DS-CDMA and MC-CDMA in the following. Refer to [10] for more discussion on MT-CDMA.

5.2 Channel and Carrier Offset Estimation

To perform multiuser detection, explicit knowledge of the effective spreading vector (at least that of user-of-interest) is required. From the mathematic viewpoint, the formulation of MC-CDMA has no es-

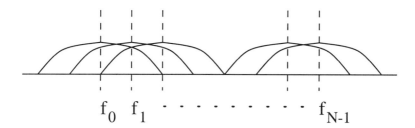

Figure 5.8 MT-CDMA spectrum.

sential difference from the regular CDMA (except for the IFFT matrix). Hence, many blind estimation approches developed for CDMA are directly applicable to MC-CDMA. For this reason we will keep our coverage of MC-CDMA parameter estimation to a minimum by only highlighting the application of the familar subspace-based techniques.

Rewrite:

$$\mathbf{a}_k = \begin{bmatrix} c_{k,1} & & 0 \\ & \ddots & \\ 0 & & c_{k,M} \end{bmatrix} \begin{bmatrix} H_k(1) \\ \vdots \\ H_k(M) \end{bmatrix} \quad (5.16)$$

and note that:

$$\begin{bmatrix} H_k(1) \\ \vdots \\ H_k(M) \end{bmatrix} = \mathbf{W} \begin{bmatrix} h_k(0) \\ \cdots \\ h_k(N_g - 1) \\ 0 \\ \vdots \\ 0 \end{bmatrix}$$

It is clear that to determine $\mathbf{a}_k$, one must estimate the channel response $\{h_k(m)\}_{i=0}^{N_g-1}$. The MUSIC-like approach described in Chapter 3 becomes readily applicable.

Given the received MC-CDMA $\mathbf{z}(n)$ in (5.4), an SVD can be performed on a collection of observation vectors to obtain the orthogonal subspace $\mathbf{U}_o$, which is orthogonal to all the effective spreading vectors:

$$\mathbf{U}_o^H \mathbf{a}_k = \mathbf{0}, \quad i = 1 \cdots P$$

Following the derivation in Chapter 3, it is not difficult to show that $\mathbf{h}_k = [h_k(0) \cdots h_k(N_g - 1)]^T$ can be uniquely identified as the least-squares solution of:

$$\mathbf{U}_o^H \begin{bmatrix} c_{k,1} & & 0 \\ & \ddots & \\ 0 & & c_{k,M} \end{bmatrix} \mathbf{W}(1:N_g)\mathbf{h}_k = \mathbf{0}$$

where $\mathbf{W}(1:N_g)$ denotes the first N_g columns of $\mathbf{W}$.

The same approach can be employed for carrier-offset estimation in flat-fading scenarios. In that case, the channel response is the Delta function and the effective spreading vector reduces to the code vector:

$$\mathbf{a}_k = \mathbf{c}_k$$

known perfectly to the receiver.

Denote ΔF_k the carrier offset of the kth user, the received MC-CDMA block is given by:

$$\mathbf{x}(n) = \begin{bmatrix} x_0(n) \\ \vdots \\ x_{M-1}(n) \end{bmatrix} = \sum_{k=1}^{P} \mathbf{\Phi}(\Delta F_k) \mathbf{W}^H \mathbf{a}_k s_{k,n} + \mathbf{v}_n \qquad (5.17)$$

where $\mathbf{\Phi}(\Delta F_k) = \text{diag}\left(1, e^{j2\pi \Delta F_k T/N}, \ldots, e^{j2\pi(N-1)\Delta F_k T/N}\right)$ is the frequency offset matrix of the kth user.

To estimate $\{\Delta F_k\}_{k=1}^{P}$ we again apply an SVD to a collection of the obversations $\{\mathbf{x}(n)\}_{n=1}^{N}$ to obtain its orthogonal subspace. It is easy to verify that for noise-free data:

$$\mathbf{U}_o^H \mathbf{\Phi}(\Delta F_k) \mathbf{W} \mathbf{c}_k = \mathbf{0}, \quad k = 1 \cdots P$$

The carrier offsets can thus be determined by evaluating the norm of the right-side vector along the unit circle.

Joint estimation of the channels and carrier offsets can in principle be accomplished using the joint two-step approach described in Chapter 3.

5.3 Enabling Codes in MC-DS-CDMA

To a great extent, the current research in this area focuses on problems concerning system architecture, modem design, and performance evaluation. In this section we will discuss detection strategies for MC-DS-CDMA and, in particular, the effect of spreading codes on receiver optimality and efficiency.

As mentioned earlier, for MC-DS-CDMA to be advantageous over MC-CDMA, MMSE detection that operates directly on the $ML \times 1$ sample vector $\mathbf{y}(k)$ has to be performed. The DC receiver, relying only on the despread samples $\mathbf{z}_i(k)$, is suboptimum in general. To quantify the performance difference, we look at the total MSEs yielded by these two methods [17]:

$$MSE_{DC} = \sum_{i=1}^{P} \left(1 - \mathbf{a}_i^H \mathbf{C}_i \mathbf{R}_{x_i x_i}^{-1} \mathbf{C}_i^H \mathbf{a}_i\right)$$
$$MSE_{MMSE} = \text{tr}\left\{\mathbf{I} - \mathbf{A}^H \mathbf{R}_{yy}^{-1} \mathbf{A}\right\} \qquad (5.18)$$

For most cases, $MSE_{MMSE} \ll MSE_{DC}$. Since the output MSEs are functions of the random channels and predesigned spreading codes, a natural question here is "Whether we can close the gap between the MSE_{MMSE} and the MSE_{DC} through judicious code selection?"

In the following, we will exploit the relationship between $\mathbf{b}_{zi}$ and $\mathbf{b}_{yi}$ and derive the enabling spreading codes that allow the simple DC receiver to achieve the performance of the more complex MMSE receiver for MC-DS-CDMA. More specifically, our goal is to derive the spreading codes that will allow a MC-DS-CDMA system to have the performance and capacity of MMSE detection with the complexity of the MC-CDMA MMSE receiver.

Note that DC reception with an $M \times 1$ combining vector $\mathbf{b}_{zi} = [b_{zi}(1) \cdots b_{zi}(M)]^T$ can be viewed as combining $\mathbf{y}(n)$ with an $ML \times 1$ *structured vector* as follows:

$$\mathbf{b}_{zi}\mathbf{z}_i = \mathbf{b}_{yi}\mathbf{y}, \quad \text{where } \mathbf{b}_{yi} = \begin{bmatrix} b_{zi}(1)\mathbf{c}_{i1} \\ \vdots \\ b_{zi}(M)\mathbf{c}_{iM} \end{bmatrix} = \mathbf{C}_i \mathbf{b}_{zi} \qquad (5.19)$$

and *vice versa* if, under certain conditions, the true MMSE receiver bears the above structure, it must be realizable using the DC receiver.

To examine this possibility, let $\mathbf{b}_{xi}^o = [b_{xi}^o(1) \cdots b_{xi}^o(M)]^T$ be the optimum combining vector for the DC receiver (i.e., it is the least-squares solution of (5.13)). Then:

$$\mathbf{C}_i^H[(\mathbf{A}\mathbf{A}^H + \mathbf{R}_v)\mathbf{C}_i\mathbf{b}_{zi}^o - \mathbf{a}_i] = 0 \tag{5.20}$$

We want to investigate its relationship with the MMSE receiver.

Denote:

$$\mathbf{r}_i = \begin{bmatrix} r_{i1} \\ \vdots \\ r_{iM} \end{bmatrix} = (\mathbf{A}\mathbf{A}^H + \mathbf{R}_n)\mathbf{C}_i\mathbf{b}_{zi}^o - \mathbf{a}_i \tag{5.21}$$

We realize from (5.15) that if $\mathbf{r}_i \equiv \mathbf{0}$, then $\mathbf{C}_i\mathbf{b}_{zi}^o$, which has the DC receiver structure, is actually the MMSE receiver because of (5.15).

The necessary and sufficient conditions for $\mathbf{r}_i \equiv \mathbf{0}$ are:

1. $\mathbf{r}_i^H\mathbf{C}_i = \mathbf{0}$

2. $\mathbf{r}_i^H\mathbf{C}_i^\perp = \mathbf{0}$

where $\mathbf{C}_i^\perp$ is the orthogonal complement of $\mathbf{C}_i$.

$\mathbf{r}_i^H\mathbf{C}_i = \mathbf{0}$ is given by (5.20). To find out the exact requirements on the spreading codes so that $\mathbf{r}_i \equiv \mathbf{0}$, we let $\mathbf{d}_i$ be an arbitrary vector in the orthogonal subspace of $\mathbf{C}_i$: $\mathbf{d}_i^H\mathbf{C}_i = \mathbf{0}$, and evaluate the product of $\mathbf{d}_i$ and $\mathbf{r}_i$: $\mathbf{d}_i^H\mathbf{r}_i = \sum_{m=1}^M \mathbf{d}_{im}^H \mathbf{r}_{im}$.

After some manipulations, we may rewrite $\mathbf{r}_{im}$ as:

$$\mathbf{r}_{im} = \sum_{l=1}^M [\sum_{k=1}^P h_{km}h_{kl}^H \mathbf{c}_{km}\mathbf{c}_{kl}^H + \sigma_{ml}]\mathbf{c}_{il}w_{xi}^o(l) - h_{im}\mathbf{c}_{im} \tag{5.22}$$

Consequently,

$$\mathbf{d}_{im}^H\mathbf{r}_{im} = \sum_{l=1}^M \left[\sum_{k=1}^P h_{km}h_{kl}^H \mathbf{d}_{im}^H\mathbf{c}_{km}\mathbf{c}_{kl}^H\mathbf{c}_{il} + \sigma_{ml}\mathbf{d}_{im}^H\mathbf{c}_{il}\right] b_{xi}^o(l) - h_{im}\mathbf{d}_{im}^H\mathbf{c}_{im}$$

$$\tag{5.23}$$

Note that $\mathbf{d}_{im}^H \mathbf{c}_{im} = 0$, thus if (1) $\mathbf{c}_{il} = \pm\mathbf{c}_{im}$, and (2) $\mathbf{c}_{im} = \pm\mathbf{c}_{jm}$ or $\mathbf{c}_{im}\mathbf{c}_{jm} = 0$,

$$\mathbf{d}_{im}^H \mathbf{r}_{im} \equiv 0, \quad \mathbf{C}_i^H \mathbf{r}_i \equiv 0 \Rightarrow \mathbf{r}_i \equiv 0. \tag{5.24}$$

The above results are summarized in the following theorem.

Theorem 2 *If users' spreading codes satisfy (1) $\mathbf{c}_{im}^H \mathbf{c}_{jm} = 0$ or $\mathbf{c}_{im} = \pm\mathbf{c}_{jm}$, and (2) $\mathbf{c}_{il} = \pm\mathbf{c}_{im}$, then the MMSE receiver and the DC receiver yield identical performance.*

Theorem 2 is a pleasant surprise. It asserts that even if orthogonality users' spreading vectors cannot be satisfied, a simple DC receiver can achieve the performance of multiuser MMSE reception at no additional cost. Further investigation reveals that the conditions in Theorem 2 are not difficult to meet in practice. One of the possible choices is the partitioned Walsh code as described below.

Corollary 1 *Let M and L be power of two. $\{\mathbf{c}_{im}\}$ satisfies all the conditions in Theorem 2 if the stacked-spreading codes, $\{\mathbf{c}_i\}$, are Walsh codes of dimension ML.*

The stacked-spreading codes $\{\mathbf{c}_i\}$ are super vectors formed by stacking the spreading codes on all subcarriers:

	subcarrier 1	$\cdots$	subcarrier M		stacked code
user 1 :	$\mathbf{c}_{11}$	$\cdots$	$\mathbf{c}_{1M}$	$\rightarrow$	$\mathbf{c}_1 = [\mathbf{c}_{11}^T \cdots \mathbf{c}_{1M}^T]^T$
$\vdots$	$\vdots$	$\vdots$	$\vdots$	$\vdots$	$\vdots$
user P :	$\mathbf{c}_{P1}$	$\cdots$	$\mathbf{c}_{PM}$	$\rightarrow$	$\mathbf{c}_P = [\mathbf{c}_{P1}^T \cdots \mathbf{c}_{PM}^T]^T$

If one chooses the Walsh codes of dimension ML (the effective spreading factor) and partitions them into the spreading codes on different subcarriers, optimum MMSE reception can be achieved by the DC receiver.

Example: Consider an 8-user, 2-subcarrier, and $L = 4$ MC-DS-CDMA system. The preceding results suggests one to construct an 8×8 Walsh matrix, partition each row into two 4-element vectors, and use them as the spreading codes for each user as shown in (5.25).

$$
\begin{array}{c|cccc|cccc}
 & \multicolumn{4}{c}{\text{subcarrier 1}} & \multicolumn{4}{c}{\text{subcarrier 2}} \\
\text{user 1:} & 1 & 1 & 1 & 1 & 1 & 1 & 1 & 1 \\
\text{user 2:} & 1 & -1 & 1 & -1 & 1 & -1 & 1 & -1 \\
\text{user 3:} & 1 & 1 & -1 & -1 & 1 & 1 & -1 & -1 \\
\text{user 4:} & 1 & -1 & -1 & 1 & 1 & -1 & -1 & 1 \\
\text{user 5:} & 1 & 1 & 1 & 1 & -1 & -1 & -1 & -1 \\
\text{user 6:} & 1 & -1 & 1 & -1 & -1 & 1 & -1 & 1 \\
\text{user 7:} & 1 & 1 & -1 & -1 & -1 & -1 & 1 & 1 \\
\text{user 8:} & 1 & -1 & -1 & 1 & -1 & 1 & 1 & -1 \\
\end{array}
\qquad (5.25)
$$

It is not a coincidence that the Walsh codes bear the structure that optimizes the DC receiver. As mentioned early, within each subcarrier the MC-DS-CDMA can be viewed as a high-density narrowband CDMA system with the number of users P greater than the spreading factor L. The partitioned Walsh codes are actually the optimum spreading sequences in the sense of meeting the Welch bound on cross correlations among users. In other words, while the spreading codes within each subcarrier can no longer be made linearly independent, the partitioned Walsh codes minimize the MAI in a high-density system. When such a structure is coordinated across all subcarriers, the result is a low-complexity receiver that offers the performance of MMSE reception. On the other hand, the general procedure to find spreading codes that satisfy the conditions in Theorem 1 when L and/or M are not integer is still under investigation.

Example: To illustrate the effect of the code design, we simulate an MC-DS-CDMA system with $M = 8$ carriers and spreading gain $L = 8$. The SNR level was set at -10dB. Random channel coefficients were used. The case compares the MMSE and DC receivers with different spreading codes. Figure 5.9 shows the output BERs of the first user when uniform codes [3] and partitioned Walsh codes are used. As seen in the top plot, the performance gap between the optimum linear receiver and the DC is significant when uniform spreading codes are used. On the other hand, when the partitioned Walsh codes are used, not only does the performance of the MMSE improve slightly, the difference between the simple DC receiver and the optimum MMSE receiver vanished as predicted in Theorem 2.

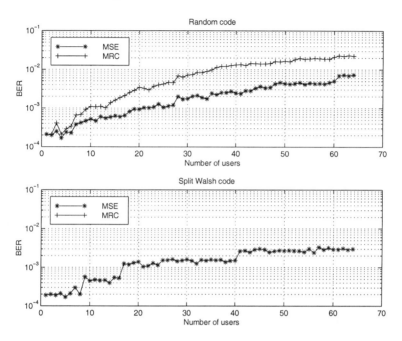

Figure 5.9 BER vs. number of users for the MMSE and DC receivers.

The above results suggest that in terms of reception complexity, the optimum linear receivers for the MC-CDMA and the MC-DS-CDMA are of the same order. The MC-DS-CDMA may be preferable to MC-CDMA in practice because of its diversities in both the time and frequency domains.

5.4 Carrier Sensitivity Analysis

In OFDM, when different symbols are transmitted through different subcarriers, the frequency dispersion manifests itself as interchannel interference, which cannot be cured except using carrier compensation at the receiver site [18]. The situation is quite different in MC-CDMA, where symbols are spread across "all" subcarriers. Interchannel interference is no longer an issue in MC-CDMA. In fact, the effect of carrier offset in MC-CDMA is very similar to that due to the multipath channels (i.e., a distortion to the effective spreading vector $\mathbf{a}_k$). For this reason, unknown carrier offsets do cause model mismatch and, consequently, performance degradation in MC-CDMA reception.

In this section, we will derive an analytical expression for the SINR degradation with frequency offset. Our analysis will focus on MC-CDMA. We assume that the transmitted symbols $s_{k,n}$ are independent and identically distributed (*i.i.d.*), and that P, the number of users, is reasonably large ($P > 10$). The interuser interference is modeled as Gaussian and the BER depends on the SINR through well-known formulas [19].

In the presence of carrier offsets, the receiver's estimate of the symbol transmitted by user j at time n is :

$$\hat{s}_{j,n} = \mathbf{b}_j^H \left(\mathbf{W} \sum_{k=0}^{K-1} \mathbf{\Phi}(\Delta F_k) \mathbf{W}^H \mathbf{a}_k s_{k,n} + \mathbf{v}_n \right) \quad (5.26)$$

We may rewrite (5.26) to separate the signal of interest from other interference,

$$\hat{s}_{j,n} = \mathbf{b}_j^H \mathbf{W} \mathbf{\Phi}(\Delta F_j) \mathbf{W}^H \mathbf{a}_j s_{j,n} +$$
$$\sum_{k=0, k \neq j}^{K-1} \mathbf{b}_j^H \mathbf{W} \mathbf{\Phi}(\Delta F_k) \mathbf{W}^H \mathbf{a}_k s_{k,n} + \mathbf{b}_j^H \mathbf{v}_n \quad (5.27)$$

In this equation, the first term is the signal of interest, the second term represents interference from other users, and the last term is Gaussian noise. Self-interference occurs through ICI, and since the same symbol is transmitted on each channel, it results in scaling the signal of interest. When the users do not experience a frequency offset, $\Delta F_k = 0$ for $k = 0, 1, \cdots, K - 1$

$$\hat{s}_{j,n} = \mathbf{b}_j^H \mathbf{a}_j s_{j,n} + \sum_{k=0, k \neq j}^{K-1} \mathbf{b}_j^H \mathbf{a}_k s_{k,n} + \mathbf{b}_j^H \mathbf{v}_n \quad (5.28)$$

Defining the vector of symbols as $\mathbf{s}_n = [s_{0,n} \cdots s_{K-1,n}]^T$ and the vector of symbol estimates at the receiver as $\hat{\mathbf{s}}_n = [\hat{s}_{0,n} \cdots \hat{s}_{K-1,n}]^T$, we find that in the absence of frequency offset, the two are related by:

$$\hat{\mathbf{s}}_n = \mathbf{B}^H \mathbf{A} \mathbf{s}_n + \mathbf{B}^H \mathbf{v}_n \quad (5.29)$$

Multicarrier CDMA

Thus, the interference between users takes on a form similar to that of DS-CDMA. Further, if each user experiences a flat-fading channel, then $\mathbf{A} = \mathbf{C}$, $\mathbf{B}^H = \mathbf{C}^H$ and thus $\hat{s}_{j,n} = s_{j,n}$ for $j = 0, 1, \cdots, K-1$. The choice of weighting vector depends on the type of detector that is implemented.

5.4.1 SINR Degradation

Let us represent the desired signal power for user j at the output of the receiver (term 1 in (5.27)) by E_j and the total signal and interference power at the output of the receiver (term 1 + term 2) by V_j. We assume that the elements of $\mathbf{v}_n$ are i.i.d. with variance σ_z^2, and $\{s_{k,n}, 0 \le k \le K-1\}$ are i.i.d. with variance σ_s^2. For notational convenience, let us define the coefficient of user k's symbol $s_{k,n}$ as:

$$\omega_{j,k} = \mathbf{b}_j^H \mathbf{W} \mathbf{\Phi}(\Delta F_k) \mathbf{W}^H \mathbf{a}_k .$$

Then the SINR at the output of the detector for user j:

$$\text{SINR}_j = \frac{E_j}{V_j - E_j + \sigma_z^2 \text{tr}\left(\mathbf{b}_j^H \mathbf{b}_j\right)} \quad (5.30)$$

$$E_j = \sigma_s^2 \, \mathcal{E}\left\{\omega_{j,j} \omega_{j,j}^*\right\} \quad (5.31)$$

$$V_j = \sigma_s^2 \, \mathcal{E}\left\{\sum_{k=0}^{K-1} \omega_{j,k} \omega_{j,k}^*\right\} \quad (5.32)$$

To obtain a simpler expression in vector notation, we define the following matrices:

$$\mathbf{X}_k = \mathcal{E}\left\{\mathbf{\Phi}(\Delta F_k) \mathbf{W}^H \mathbf{a}_k \mathbf{a}_k^H \mathbf{W} \mathbf{\Phi}(\Delta F_k)^H\right\}$$

$$\mathbf{X} = \sum_{k=0}^{K-1} \mathbf{X}_k$$

where the expectation is over the random variables ΔF_j. Thus, the expressions for desired signal power and total signal and interference power for user j, in vector notation, are:

$$E_j = \sigma_s^2 \left(\mathbf{b}_j^H \mathbf{W} \mathbf{X}_j \mathbf{W}^H \mathbf{b}_j\right) \quad (5.33)$$

$$V_j = \sigma_s^2 \left(\mathbf{b}_j^H \mathbf{W} \mathbf{X} \mathbf{W}^H \mathbf{b}_j\right) \quad (5.34)$$

To simplify this expression, we use the following property:

Lemma 1: Let $\mathbf{A} \in \mathcal{C}^{N \times N}$ and let $\mathbf{D} \in \mathcal{C}^{N \times N}$ be a diagonal matrix with nonzero elements $\mathbf{D} = \text{diag}\,[d_0, d_1, \cdots, d_{N-1}]$. Then:

$$\mathbf{D}\mathbf{A}\mathbf{D}^H = \left(\mathbf{d}\mathbf{d}^H\right) \odot \mathbf{A} \qquad (5.35)$$

where $\mathbf{d} = [d_0, d_1, \cdots, d_{N-1}]^T$.

Let us define a vector:

$$\mathbf{p}(\Delta F_k) = \left[1, e^{j2\pi \Delta F_k T/N}, \cdots, e^{j2(N-1)\pi \Delta F_k T/N}\right]^T$$

composed of the diagonal elements of matrix $\mathbf{\Phi}(\Delta F_k)$. Then $\mathbf{X}$ can be rewritten as follows:

$$\mathbf{X} = \sum_{k=0}^{K-1} \mathcal{E}\left\{\mathbf{p}(\Delta F_k)\mathbf{p}(\Delta F_k)^H\right\} \odot \left(\mathbf{W}^H \mathbf{a}_k \mathbf{a}_k^H \mathbf{W}\right)$$

If the random variables ΔF_k are identically distributed, then the matrix $\mathcal{E}\left\{\mathbf{p}(\Delta F_k)\mathbf{p}(\Delta F_k)^H\right\}$ is the same for each user k, and:

$$E_j = \sigma_s^2 \,\text{tr}\left[\left(\mathbf{\Psi} \odot \left(\mathbf{W}^H \mathbf{a}_j \mathbf{a}_j^H \mathbf{W}\right)\right) \mathbf{W}^H \mathbf{b}_j \mathbf{b}_j^H \mathbf{W}\right] \qquad (5.36)$$

$$V_j = \sigma_s^2 \,\text{tr}\left[\left(\mathbf{\Psi} \odot \sum_{k=0}^{K-1}\left(\mathbf{W}^H \mathbf{a}_k \mathbf{a}_k^H \mathbf{W}\right)\right) \mathbf{W}^H \mathbf{b}_j \mathbf{b}_j^H \mathbf{W}\right] \qquad (5.37)$$

$$= \sigma_s^2 \,\text{tr}\left[\left(\mathbf{\Psi} \odot \left(\mathbf{W}^H \mathbf{A}\mathbf{A}^H \mathbf{W}\right)\right) \mathbf{W}^H \mathbf{b}_j \mathbf{b}_j^H \mathbf{W}\right]$$

where $\mathbf{\Psi} = \mathcal{E}\left\{\mathbf{p}(\Delta F_0)\mathbf{p}(\Delta F_0)^H\right\}$ has elements:

$$\Psi_{k,l} = \mathcal{E}\left\{\exp(j2\pi(k-l)\Delta F_0 T/N)\right\}$$

This expression depends on the probability distribution of ΔF_0, but it is easy to calculate, since it has the same form as the characteristic function of ΔF_0.

The calculation of E_j and V_j can be further simplified if we note the following property,

Lemma 2: Let $\mathbf{A}, \mathbf{B} \in \mathcal{C}^{N \times N}$. Then:

$$\text{tr}(\mathbf{A}\mathbf{B}^T) = \text{sum}\,(\mathbf{A} \odot \mathbf{B}) \qquad (5.38)$$

Multicarrier CDMA

Rearranging (5.38) and applying this lemma yields the following more compact result :

Result 1: Given the system described in the previous section, assuming that the transmitted symbols are *i.i.d.* with variance σ_s^2, and that the additive white Guassian noise (AWGN) is also *i.i.d.* with variance σ_z^2, the SINR for user k at the detector output is:

$$\text{SINR}_j = \frac{E_j}{V_j - E_j + \frac{\sigma_z^2}{\sigma_s^2}\left(\mathbf{b}_j^H \mathbf{b}_j\right)} \quad (5.39)$$

$$E_j = \text{sum}\left(\boldsymbol{\Psi} \odot \left(\mathbf{W}^H \mathbf{a}_j \mathbf{a}_j^H \mathbf{W}\right) \odot \left(\mathbf{W}^H \mathbf{b}_j \mathbf{b}_j^H \mathbf{W}\right)^*\right) \quad (5.40)$$

$$V_j = \text{sum}\left(\boldsymbol{\Psi} \odot \left(\mathbf{W}^H \mathbf{A} \mathbf{A}^H \mathbf{W}\right) \odot \left(\mathbf{W}^H \mathbf{b}_j \mathbf{b}_j^H \mathbf{W}\right)^*\right) \quad (5.41)$$

Hence, the degradation in SINR caused by frequency offset, D_j, is given by:

$$D_j = \frac{\text{SINR}_j}{\text{SINR}_j^0} \quad (5.42)$$

where:

$$\text{SINR}_j^0 = \frac{\mathbf{b}_j^H \mathbf{a}_j \mathbf{a}_j^H \mathbf{b}_j}{\mathbf{b}_j^H (\mathbf{A}\mathbf{A}^H - \mathbf{a}_j \mathbf{a}_j^H + \frac{\sigma_z^2}{\sigma_s^2}\mathbf{I})\mathbf{b}_j} \quad (5.43)$$

is the average SINR for user j in a MC-CDMA system with no frequency-offset error.

We have studied this expression for two particular distributions of the frequency-offset ΔF_0.

- If ΔF_0 is uniformly distributed in the interval $[-F, F]$, we find:

$$\begin{aligned}\mathcal{E}\left\{\exp(j 2\pi(k-l)\Delta F_0 T/N)\right\} &= \text{sinc}(2\pi(k-l)FT/N) \\ &= \text{sinc}(2\pi(k-l)\sqrt{3}\sigma_F T/N)\end{aligned} \quad (5.44)$$

- If ΔF_0 is Gaussian with zero mean and variance σ_F,

$$\begin{aligned}\mathcal{E}\left\{\exp(j 2\pi(k-l)\Delta F_0 T/N)\right\} &= \exp(-(2\pi(k-l)\sigma_F T/N)^2/2)\end{aligned} \quad (5.45)$$

Both functions in (5.44) and (5.45), $\text{sinc}(2\pi(k-l)\sqrt{3}\sigma_F T/N)$ and $\exp(-(2\pi(k-l)\sigma_F T/N)^2/2)$, are approximated quite well by $\cos(2\pi(k-l)\sigma_F T/N)$ for $\sigma_F(k-l) < 1$. Even when σ_F is relatively large, there are few terms in the matrix $\mathbf{\Psi}$, where $k - l$ is large enough that the approximation does not hold. Thus, we find that for both uniform-distributed and Gaussian distributed ΔF_0,

$$\begin{aligned}\Psi_{k,l} &= \mathcal{E}\left\{\exp\left(2\pi(k-l)\Delta F_0 T/N\right)\right\} \\ &\approx \cos(2\pi(k-l)\sigma_F T/N) \\ &= \Re\left\{\exp(j2\pi(k-l)\sigma_F T/N)\right\}\end{aligned} \quad (5.46)$$

Figure 5.10 shows plots of the three functions versus intercarrier distance, $(k-l)/N$, for $\sigma_F T = 0.2$. Although this is quite a large offset, the plots are still quite close.

Taking into account the Hermitian structure of the matrices in (5.40) and (5.41), we find that:

$$\begin{aligned}&\text{sum}\left(\mathbf{\Psi}\odot\left(\mathbf{W}^H\mathbf{A}\mathbf{A}^H\mathbf{W}\right)\odot\left(\mathbf{W}^H\mathbf{b}_j\mathbf{b}_j^H\mathbf{W}\right)^*\right)\\ &= \text{sum}\left(\Re\{\hat{\mathbf{\Psi}}\}\odot\left(\mathbf{W}^H\mathbf{A}\mathbf{A}^H\mathbf{W}\right)\odot\left(\mathbf{W}^H\mathbf{b}_j\mathbf{b}_j^H\mathbf{W}\right)^*\right)\\ &= \frac{1}{2}\text{sum}\left(\hat{\mathbf{\Psi}}\odot\left(\mathbf{W}^H\mathbf{A}\mathbf{A}^H\mathbf{W}\right)\odot\left(\mathbf{W}^H\mathbf{b}_j\mathbf{b}_j^H\mathbf{W}\right)^*\right)\\ &+ \frac{1}{2}\text{sum}\left(\hat{\mathbf{\Psi}}^*\odot\left(\mathbf{W}^H\mathbf{A}\mathbf{A}^H\mathbf{W}\right)\odot\left(\mathbf{W}^H\mathbf{b}_j\mathbf{b}_j^H\mathbf{W}\right)^*\right)\end{aligned}$$

where $\hat{\mathbf{\Psi}}$ is a matrix such that $\mathbf{\Psi} = \Re\{\hat{\mathbf{\Psi}}\}$. In particular, since $\Psi_{k,l} = \cos(2\pi(k-l)\sigma_F T/N)$, we may choose $\hat{\Psi}_{k,l} = \exp(j2\pi(k-l)\sigma_F T/N)$, so that $\hat{\mathbf{\Psi}} = \mathbf{p}(\sigma_F)\cdot\mathbf{p}(\sigma_F)^H$

$$\begin{aligned}&\text{sum}\left(\mathbf{\Psi}\odot\left(\mathbf{W}^H\mathbf{A}\mathbf{A}^H\mathbf{W}\right)\odot\left(\mathbf{W}^H\mathbf{b}_j\mathbf{b}_j^H\mathbf{W}\right)^*\right)\\ &= \frac{1}{2}\text{sum}\left(\mathbf{p}(\sigma_F)\mathbf{p}(\sigma_F)^H\odot\left(\mathbf{W}^H\mathbf{A}\mathbf{A}^H\mathbf{W}\right)\odot\left(\mathbf{W}^H\mathbf{b}_j\mathbf{b}_j^H\mathbf{W}\right)^*\right)\\ &+ \frac{1}{2}\text{sum}\left(\mathbf{p}(-\sigma_F)\mathbf{p}(-\sigma_F)^H\odot\left(\mathbf{W}^H\mathbf{A}\mathbf{A}^H\mathbf{W}\right)\odot\left(\mathbf{W}^H\mathbf{b}_j\mathbf{b}_j^H\mathbf{W}\right)^*\right)\\ &\approx \frac{1}{2}\left(\mathbf{b}_j^H\mathbf{W}\mathbf{\Phi}(\sigma_F)\mathbf{W}^H\mathbf{A}\mathbf{A}^H\mathbf{W}\mathbf{\Phi}(\sigma_F)^H\mathbf{W}^H\mathbf{b}_j\right)\\ &+ \frac{1}{2}\left(\mathbf{b}_j^H\mathbf{W}\mathbf{\Phi}(-\sigma_F)\mathbf{W}^H\mathbf{A}\mathbf{A}^H\mathbf{W}\mathbf{\Phi}(-\sigma_F)^H\mathbf{W}^H\mathbf{b}_j\right)\end{aligned}$$

We conclude that even though in the uplink each user experiences a different frequency offset, if these offsets are distributed identically as above and the frequency offset is reasonably small ($\sigma_F T \leq 0.2$), then the average SINR can be expressed as a function of the standard deviation of the frequency offset only.

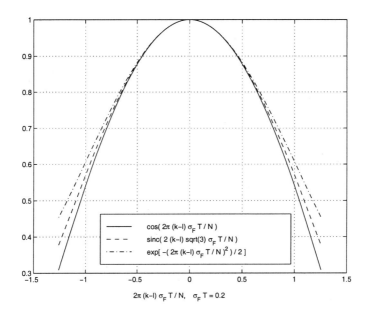

Figure 5.10 Plot of $\Psi_{k,l}$ for $\sigma_F = 0.2$.

In view of this result, (5.31) and (5.32) can be used to generate a more compact expression for the desired signal power E_j and the total signal and interference power V_j. Let us define the matrix $\mathbf{\Omega}(\sigma_F) = \mathbf{B}^H \mathbf{W} \mathbf{\Phi}(\sigma_F) \mathbf{W}^H \mathbf{A}$, whose rows are the vectors $\mathbf{\Omega}_j(\sigma_F) = \mathbf{b}_j^H \mathbf{W} \mathbf{\Phi}(\sigma_F) \mathbf{W}^H \mathbf{A}$, $0 \leq j \leq K-1$, and whose elements are $\omega_{j,k}(\sigma_F) = \mathbf{b}_j^H \mathbf{W} \mathbf{\Phi}(\sigma_F) \mathbf{W}^H \mathbf{a}_k$. Then:

$$E_j = \frac{1}{2} \left(|\omega_{j,j}(\sigma_F)|^2 + |\omega_{j,j}(-\sigma_F)|^2 \right) \quad (5.47)$$

$$V_j = \frac{1}{2} \sum_{k=0}^{K-1} \omega_{j,k}(\sigma_F) \omega_{j,k}^*(\sigma_F) + \omega_{j,k}(-\sigma_F) \omega_{j,k}^*(-\sigma_F) \quad (5.48)$$

148 *Signal Processing Applications in CDMA Communications*

$$= \frac{1}{2}\left(\|\mathbf{\Omega}_j(\sigma_F)\|^2 + \|\mathbf{\Omega}_j(-\sigma_F)\|^2\right)$$

It is possible to write such an expression because $\mathbf{\Psi}$ has been shown to be well-approximated by the outer product of a column matrix, as shown in (5.47) and (5.48). The SINR can be calculated from (5.39) by substituting these values.

From the approximation:

$$\mathbf{\Psi} \approx 0.5\left(\mathbf{p}(\sigma_F)\mathbf{p}(\sigma_F)^H + \mathbf{p}(-\sigma_F)\mathbf{p}(-\sigma_F)^H\right)$$

we observe that when the standard deviation of frequency offset increases, the off-diagonal terms of $\mathbf{\Psi}$ will decrease while the diagonal terms stay the same. Thus, the user will experience more self-interference, and the power received from the desired user will be less. Since the frequency offset will have the same effect on the interuser interference, the total received power will also decrease. Thus, in theory, the net effect of the frequency offset is to make the AWGN much more powerful. This analysis shows the limitations of the approximation (5.46), but it does not take into account the inter-user interference resulting from having different frequency offsets for different users. Any user, then, who has a smaller frequency offset than the average will improve his own error but will degrade the SINR for all other users, causing a net decrease in SINR. The error in our prediction becomes significant as σ_F increases, particularly for systems for which the inter-user interference is very low in the absence of frequency offset. For large σ_F, this approximation will overestimate the actual SINR.

5.5 Discussion

In this section we verify our results with simulations. We also discuss how various factors such as the detection method (coherent combining receiver (CCR) or MMSE), the number of users, and the subcarrier spacing affect the sensitivity of the system to frequency-offset error.

Outline of the Simulated System:

Two systems are studied: a tightly packed, OFDM-like system (OFDM-CDMA, Figure 5.5(a)), and one where every other available

Multicarrier CDMA 149

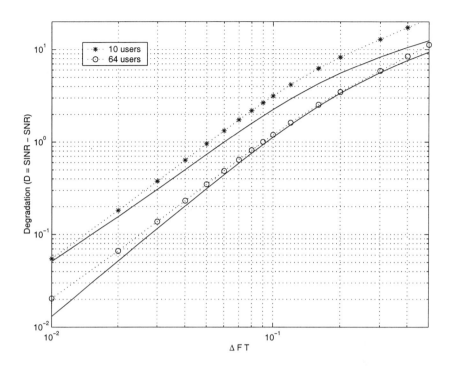

Figure 5.11 Verifying (5.42) for an MMSE receiver: The dotted lines are simulations of MPFB-CDMA with $M = 64$, $\beta = 2$, $K = 10, 64$. The solid lines are corresponding plots of (5.42).

subchannel is used (MPFB-CDMA, Figure 5.5(b)). We refer to the second system as a multiphase filterbank (MPFB), since a filterbank structure is used to implement the transmitter and receiver filters. When the distance between subcarriers is increased by an integer multiple of $1/T$, the resulting tones are orthogonal and filters do not affect the error response to frequency offset significantly. To achieve a fair comparison, both systems are given the same bandwidth, thus the OFDM system has 128 carriers while the MPFB has 64. The total power of each user is set to 1. DQPSK is used in both systems. The frequency offsets are chosen from a zero-mean Gaussian distribution. In this section, the degradation in SINR relative to the case where there is no frequency offset is plotted against the standard de-

150 *Signal Processing Applications in CDMA Communications*

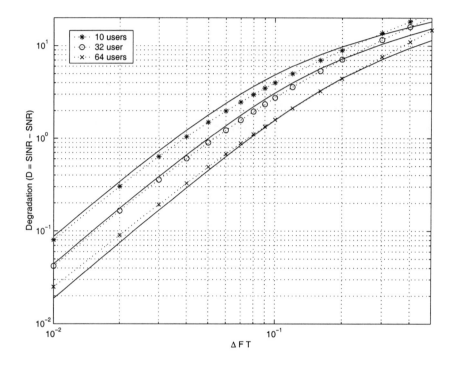

Figure 5.12 Verifying (5.42) for an MMSE receiver: The dotted lines are simulations of OFDM-CDMA with $N = M = 128$, $\beta = 1$, $K = 32, 64, 128$. The solid lines are corresponding plots of (5.42).

viation of the frequency offset. In Figures (5.11) to (5.14), the results are presented for individual channel vectors. In all other figures, the degradation is averaged over an ensemble of channels.

Accuracy of the Estimate:

Let us first examine the accuracy of the result derived for the two schemes outlined above, OFDM and MPFB. In Figures 5.11 to 5.14, we have plotted the expression for the degradation in SINR obtained from (5.42) against simulation results. Figure 5.11 shows the result for MPFB-CDMA ($\beta = 2$) with an MMSE receiver, Figure 5.12 shows the result for OFDM-CDMA with an MMSE receiver, Figure 5.13 shows the result for MPFB-CDMA ($\beta = 2$) with a coherent combining

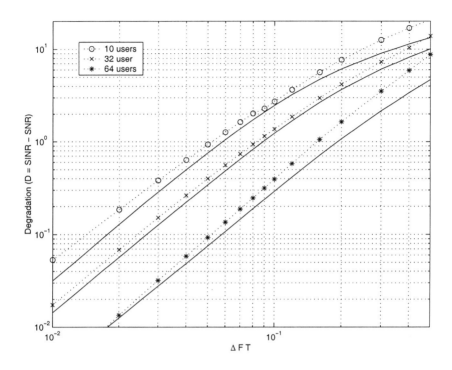

Figure 5.13 Verifying (5.42) for a CCR receiver: The dotted lines are simulations of MPFB-CDMA with $M = 64$, $\beta = 2$, $K = 32, 64$. The solid lines are corresponding plots of (5.42).

receiver, and Figure 5.14 shows the result for OFDM-CDMA with a coherent combining receiver. As expected, the match is quite close in all cases. The figures also depict the sensitivity of both systems to changes in the number of users. Fewer users means less intersymbol interference.

Single User Detection versus Multiuser Detection:

Figures 5.15 and 5.16 compare the CCR to the MMSE receiver for MPFB-CDMA. In this example, we used channel coefficients with mean 1 and standard deviation 0.5, so that we could obtain meaningful results from the CCR. While the CCR's performance does not degrade as quickly as the MMSE receiver's performance, the CCR is

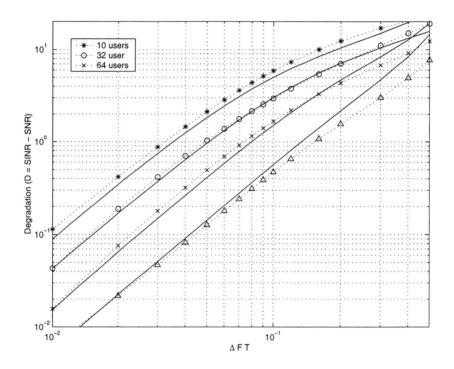

Figure 5.14 Verifying (5.42) for a CCR receiver: The dotted lines are simulations of OFDM-CDMA with $N = M = 128$, $\beta = 1$, $K = 32, 64, 128$. The solid lines are corresponding plots of (5.42).

already doing very badly. When there is no frequency offset, the SNR at the output of the CCR with 64 users is 5.8dB while that of the MMSE multiuser detector is 14.7dB. However, the sensitivity of the MMSE receiver to unknown frequency offset suggests that with imperfect knowledge of the channel, using the CCR may be a viable option.

CDMA Coding Gain:

Since CDMA is employed, if the number of users is less than the coding gain of the system, the signal-to-noise ratio at the receiver output will be better than that on each channel at the input. The higher the initial SNR, the more it will degrade with frequency offset. A low-

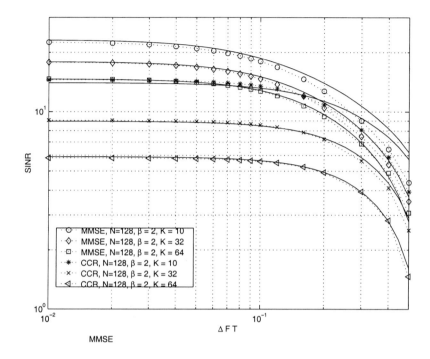

Figure 5.15 Simulation results, coherent combining receiver versus MMSE receiver.

initial SNR means that AWGN, which is not sensitive to frequency offset, dominates the noise term, hence noise remains uniformly high as σ_F varies. In Figures 5.11 to 5.14, we see that when there are fewer users, there is more degradation in SINR.

Changing the Number of Subcarriers:

The effect of an increase in the number of subcarriers depends on where we get the subcarriers. If we change the number of subcarriers, while keeping N/K and T constant, then the degradation in SINR will not change. However, if the overall signaling rate for each user is kept constant while more tones are added, then the new system will be more sensitive to frequency offset. In this situation, the same total bandwidth is used to accommodate more subcarriers. The spacing between subcarriers is narrower but because the symbols are longer, the

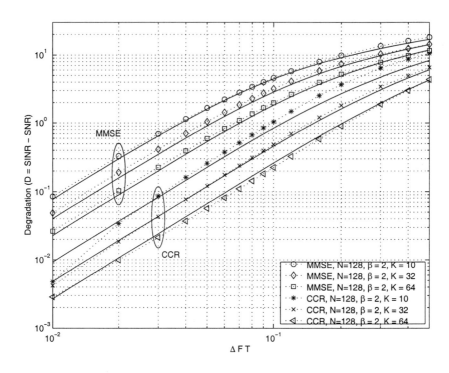

Figure 5.16 Simulation results, coherent combining receiver versus MMSE receiver.

system is more robust to multipath. Figure 5.17 illustrates this. The points marked "*" come from a simulation with $N_1 = 128$ available subcarriers, $M_1 = 64$ subcarriers used, $\beta_1 = 2$, $K_1 = 64$ users, and total bandwidth $B_1 = N_1/(2T)$. The points marked "o" are from a simulation with $N_2 = 256$, $M_2 = 128$, $K_2 = 128$ and total bandwidth $B_2 = N_2/(2T) = 2B_1$. As we can see, the two plots coincide. The last simulation is of a system with $M_3 = 128$ subcarriers used out of $N_3 = 256$ available and $K_3 = 64$ users, so that $T_3 = 2T$. The total bandwidth of this system is $B_3 = B_1$, but the same frequency offset, normalized to T, causes greater degradation for this system, where subcarriers are placed closer together.

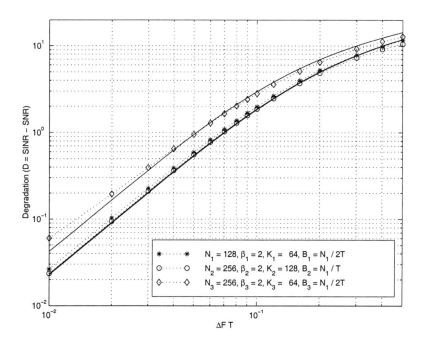

Figure 5.17 Illustrating how frequency-offset error changes when the number of subcarriers is varied. Dotted lines are simulation results, solid lines are from (5.42).

Interspersing Subcarriers:

From the expression in (5.40), it is clear that increasing β, the distance between the utilized carriers, will reduce the ICI. For this reason, when there are fewer users than subcarriers, we may take advantage of this by letting β be greater than 1 and using only every βth subcarrier to transmit on. We verify by simulations that this is useful: comparing the plots in Figure 5.18, we see that for a given number of users, at a given frequency offset, OFDM-CDMA suffers more degradation than MPFB-CDMA. Thus, when the number of carriers is fixed and the number of users is fewer than half the number of available carriers, it is advantageous to use fewer carriers that are spaced further apart.

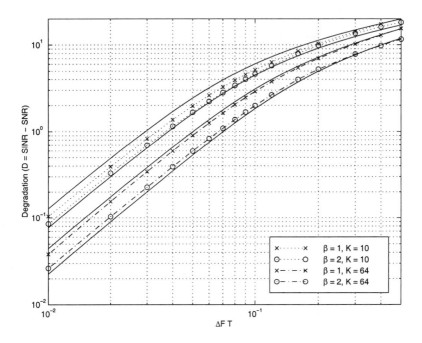

Figure 5.18 Illustrating how frequency-offset error changes when the number of subcarriers is varied, $N = 128$ in each case. The number of information-bearing subcarriers and the number of users are varied. Dotted lines are simulation results, solid lines are from (5.42).

5.6 Conclusion

In summary, MC-CDMA that spreads and modulates data stream onto parallel subcarriers has the advantages of both CDMA and MCM in terms of narrowband fading/interference resistance and system flexibility. At the same time, it also suffers from increased sensitivity to frequency offset and nonlinear amplifications. The analysis in this chapter shows that multiuser detection can be made feasible by judicious design of the spreading codes and key system parameters, such as the number of subcarriers and the guard interval. On the other hand, these high-performance detectors are more sensitive to carrier impairment. In the presence of unknown frequency offset, the performance of MMSE multiuser receivers degrade rapidly, suggesting that with moderate offset, single-user receivers may provide a better trade-off between complexity and performance.

References

[1] N. Yee, J. P. Linnartz, and G. Fettweis, "Multicarrier CDMA in indoor wireless radio," In *Proc. IEEE MILCOM*, pages 52–56, Boston, MA, October 1993.

[2] L. Tomba and W. A. Krzymien, "Downlink detection schemes for MC-CDMA systems in indoor environments," *IEICE Trans. Communications*, **E79-B**(9):1351–1360, September 1996.

[3] S. Kondo and L. B. Milstein, "Performance of multicarrier DS-CDMA systems," *IEEE Trans. on Communications*, **44**(2):238–246, February 1996.

[4] E. A. Sourour and M. Nakagawa, "Performance of orthogonal multicarrier CDMA in a multipath fading channel," *IEEE Trans. on Communications*, **44**(3):356–367, March 1996.

[5] Q. Chen, E. S. Sousa, and S. Pasupathy, "Multicarrier CDMA with adaptive frequency-hopping for mobile radio systems," *IEEE Journ. on Selected Areas in Communication*, **14**(9):1852–1858, December 1996.

[6] L. Vandendorpe, "Multitone spread spectrum multiple-access communications system in a multipath Rician fading channel," *IEEE Trans. on Vehicular Technology*, **44**(2):327–337, May 1995.

[7] R. Prasad and S. Hara, "An overview of multicarrier CDMA," In *Proc. IEEE ISSSTA*, pages 107–114, Mainz, Germany, September 1996.

[8] A. Chouly, A. Brajal, and S. Jourdan, "Orthogonal multicarrier techniques applied to direct sequence spread spectrum CDMA system," In *Proc. IEEE Globecom*, pages 1723–1728, Houston, USA, November 1993.

[9] J. Bingham. "Multicarrier modulation for data transmission: An

idea whose time has come," *IEEE Communications Magazine*, 28(5):982–989, May 1990.

[10] L. Vandendorpe, "Multitone spread spectrum communication systems in a multipath Rician fading channel," In *Proceedings of the International Zurich Seminar on Digital Communications*, page 440, Zurich, Switzerland, March 1994. Lecture notes in Computer Science Mobile Communications Advanced Systems and Components, Springer-Verlag.

[11] V. M. DaSilva and E. S. Sousa, "Performance of orthogonal CDMA codes for quasi-synchronous communication systems," In *Proceedings of the IEEE International Conference on Universal Personal Communications*, pages 995–99, Ottowa, Canada, October 1993.

[12] S. Kondo and L. B. Milstein, "Performance of multicarrier DS-CDMA systems," *IEEE Trans. on Communications*, **44**(2):238–46, February 1996.

[13] A. Brajal, A. Chouly, and S. Jourdan, "Orthogonal multicarrier techniques applied to direct-sequence spread spectrum CDMA systems," In *Proceedings of the IEEE Global Telecommunications Conference*, pages 1723–28, Houston, TX, November 1993.

[14] N. Yee and J. P. Linnartz, "BER of multicarrier CDMA in an indoor Rician fading channel," In *Proceedings of the Asilomar Conference on Signals, Systems, and Computers*, Volume 1, pages 426–30, Pacific Grove, CA, November 1993.

[15] K. Fazel and L. Papke, "On the performance of convolutionally-coded CDMA/OFDM for mobile communication systems," In *Proceedings of the International Symposium on Personal, Indoor, and Mobile Radio Communications*, pages 468–72, Yokohama, Japan, September 1993.

[16] S. Hara and R. Prasad, "Overview of multicarrier CDMA," *IEEE Communications Magazine*, **35**(12):126–33, December 1997.

[17] B. Ottersten, R. Roy, and T. Kailath, "Signal waveform estimation in sensor array processing," In *Proc. 23rd Asilomar Conference on Signals, Systems, and Computers*, Volume 2, pages 787–791, Pacific Grove, California, November 1989.

[18] J. C. Chuang, W. Rhee, and Jr L. J. Cimini, "Performance com-

parison of OFDM and multitone with polyphase filter bank for wireless communications," In *Proceedings of the Vehicular Technology Conference, Part 2 (of 3)*, pages 768–72, Ottowa, Ontario, Canada, May 1998.

[19] M. Van-Bladel, T. Pollet, and M. Moeneclaey, "BER sensitivity of OFDM systems to carrier frequency offset and Wiener phase noise," *IEEE Trans. on Communications*, **43**(2–4):191–93, February to April 1995.

Chapter 6

Space-Division Multiple-Access

One of the distinctive features of CDMA is that it is both a physical layer modulation and a medium access control (MAC) layer multiple-access scheme. Traditional design paradigms for wireless networks are mostly layered, allowing little or no information exchange between the link layer and the physical layer. As a result, the role of signal processing has been primarily limited to the physical layer, and solving problems typically concerns signal enhancement, noise/interference reduction, and parameter estimation. As wireless services evolve to meet increasing end-user expectations, the interactions between layers that have not been sufficiently explored become increasingly important. Some of the cross-layer problems, ranging from architectural principles for multimedia communications to scalable multirate CDMA, can be approached from signal processing perspectives.

In the last chapter of this book, we will discuss an entirely different type of application of signal processing in wireless communication. In particular, we will describe a packet radio network that exploits the spatial diversity provided by antenna arrays at both the physical and MAC layers. The access mode considered combines CDMA (including the extreme cases of TDMA and FDMA; see Chapter 1) with the so-called space-division multiple-access (SDMA). The MAC protocol is "channel-aware" in the sense that it adapts to the channel conditions of an antenna array system to achieve throughput multiplication and reduction of packet delays. Such gain, provided by a unique cross-layer design, cannot be obtained by physical layer processing alone. We will show that the bottom-up design has the potential to make a substantial impact on future applications such as wireless LANs.

6.1 Background

Many innovative MAC protocols have been proposed in the past decade; see [1, 2, 3]. Most wireless data networks inherit (at least partially) the architecture of wireline systems and treat "layers" as largely uncoupled entities [4, 5]. For example, data link/media-access protocols have been frequently designed and analyzed without sufficient regard for design details at the physical layer. Furthermore, underlying most work in improved protocols is the assumption that wireless resources (time, frequency, codes) can be reliably managed at the base station, whereas in reality radio channels often experience large variations due to mobility and propagation attenuation.

One of the key resources in wireless communication is the spatial diversity provided by antenna arrays. It is well-known that such diversity, when used properly through array processing, can effectively combat fading/jamming/interference and significantly increase the capacity of a wireless system. While inherently rich, the space resource is highly "irregular" and thus is difficult to deal with at the network level. For this reason, exploitation of the spatial diversity has been limited to the physical layer using techniques such as spatial beamforming. However, if the MAC protocol is designed without adaptability to the channel conditions, it has to be designed for the "worst-case" scenario and, as a result, there will be too much signal level margin and a significant waste of network resources.

The need for coordinated MAC layer and physical layer to a system architecture is pronounced in antenna array networks. Only limited results are available to date, one of which is the SDMA scheme that attempts to multiply the throughput of wireless systems by introducing spatial channels [6, 7, 8]. For a system with M antenna elements, the underlying idea is to divide each slot (code/time/frequency) into M "spatial slots" so that the total number of noninterfering slots is increased by M folds.

While intuitively promising, an obvious flaw of this scheme is that spatial channels are rarely orthogonal in practice (even with optimum detection). If multiple users are assigned to one slot without considering these spatial characteristics, the ones with unfavorable spatial configuration will experience significant throughput disadvantages. Since the effectiveness of spatial separation depends on the base station array responses (often referred to as the spatial signatures) of all co-slot users, the instantaneous signal-to-noise ratio (SNR) of beamforming outputs can vary dramatically.

A fundamental solution to the above problem is the "channel-aware" protocol that controls the traffic based on the spatial characteristics of the terminals. In SDMA in particular, the performance of the system can be enhanced with spatial signature-based scheduling (e.g., assigning "most orthogonal" terminals to the same time slot to increase the traffic throughput) and load-dependent slot adjustment. Such a MAC treatment allows a system to exploit the spatial diversity in an efficient manner with fixed-complexity physical-layer processing.

The main focus of this chapter is a scheduling protocol for slotted antenna array packet networks. We present a scheme to increase the SDMA throughput by allocating terminals into slots of variable length based on these spatial characteristics. System adaptation is carried out at the MAC layer, and the computationally demanding operations can be implemented using high-efficiency dedicated circuitry. We show that with the new scheme, a much more substantial improvement can be obtained than with the regular SDMA. The efficiency of SDMA scheduling is demonstrated by comparing its throughput with the SDMA throughput upper-bound and lower-bound derived in the book. Another salient feature of the new protocol is that it has a minimum performance gap between the average and worst case throughput of the terminals, indicating maximum capacity improvement in worst-case limited applications. Furthermore, we analyze the behavior of the protocol for a large population of terminals and show its asymptotic optimality.

164 Signal Processing Applications in CDMA Communications

The remainder of this chapter is organized as follows. In Section 6.2, we provide an overview of slotted antenna array packet networks and the conventional SDMA. Section 6.3 describes the scheduling protocol, including a slot allocation scheme and a mechanism for slot adjustment. The performance of the protocol is analyzed in Section 6.4 where we derive the throughput upper- and lower-bounds and the asymptotic performance of the protocol. The analytic results are verified through computer simulations in Section 6.5.

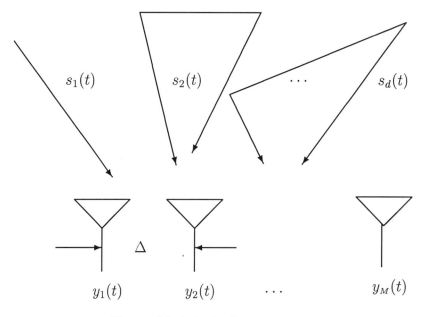

Figure 6.1 A typical array scene.

6.2 Antenna Array Packet Networks

Antenna array processing has been an active topic for wireless communications for the last 10 years. Through controlling its beam pattern, an antenna array can improve some key operation parameters such as the SINR over a single antenna wireless system. Figure 6.1 depicts a typical antenna array system where the spatial diversity is exploited

Space-Division Multiple-Access   165

for multiple-access communications. In the absence of noise, the response of an M-element antenna array to a narrowband source $s(t)$ can be written as:

$$\mathbf{y}(t) \overset{\text{def}}{=} [y_1(t)\ y_2(t)\ \cdots\ y_M(t)]^T = \mathbf{a}s(t)$$

where $(\cdot)^T$ denotes transposition, and $\mathbf{a} = [a(1)\ a(2)\ \cdots\ a(M)]^T$ is the array response vector that captures the spatial characteristics (direction-of-arrivals, number of multipath reflections, and attenuation) of the terminal. In some literature, the array response vector is referred to as the *spatial signature*.

When d terminals communicate simultaneously with the base station, the total output of the antenna array is given by:

$$\mathbf{y}(t) = \sum_{k=1}^{d} \mathbf{a}_k s_k(t) + \mathbf{n}(t) = \mathbf{A}\mathbf{s}(t) + \mathbf{n}(t) \quad (6.1)$$

$$\mathbf{A} = [\mathbf{a}_1, \ldots, \mathbf{a}_d],\ \mathbf{s}(t) = [s_1(t), \ldots, s_d(t)]^T$$

$$\mathbf{n}(t) = [n_1(t),\ n_2(t),\ \cdots, n_M(t)]^T \quad (6.2)$$

where $\mathbf{n}(t)$ is the additive noise vector and $\mathbf{A}$ is defined as the "array manifold" whose columns are the spatial signatures. To facilitate our presentation in the remainder of this book, we assume (1) all signals have unit power and $\|\mathbf{a}_k\| = 1$ (i.e., perfect power control), and (2) the noise is i.i.d. with strength σ_n^2. Under these assumptions, the covariance matrix of the array output has the form of:

$$\mathbf{R_{yy}} = E\{\mathbf{y}(t)\mathbf{y}^H(t)\} = \mathbf{A}\mathbf{A}^H + \sigma_n^2 \mathbf{I} \quad (6.3)$$

In most practical situations, the spatial signatures for different terminals are different, allowing the base station to decouple superimposed signals through spatial separation. To retrieve individual signals (e.g., $s_k(t)$, from the antenna outputs), outputs of the antenna array are weighted-and-summed with a set of weight coefficients:

$$\hat{s}_i(t) = \sum_{m=1}^{M} w_i^*(m) y_m(t)$$

with $\{w_i(m)\}$ designed so as to constructively combine the signal-of-interest and destructively combine the interference-plus-noise. This processing is referred to as "spatial beamforming" [9]. The optimum weight vector $\mathbf{w}_k = [w_k(1) \cdots w_k(M)]^T$ that minimizes the mean-squared error (MSE) of the signal estimate is given by:

$$\mathbf{w}_k = \mathbf{R}_{\mathbf{yy}}^{-1} \mathbf{a}_k$$

in which case the output MSE and SINR, respectively, of the kth signal estimate:

$$MSE_k = 1 - \mathbf{a}_i^H \mathbf{R}_y^{-1} \mathbf{a}_i \qquad (6.4)$$

$$SNR_k = \frac{\|\mathbf{a}_k^H \mathbf{R}_{\mathbf{yy}}^{-1} \mathbf{a}_k\|^2}{\sum_{j \neq k}^{K} \|\mathbf{a}_j^H \mathbf{R}_{\mathbf{yy}}^{-1} \mathbf{a}_k\|^2 + \|\mathbf{a}_k^H \mathbf{R}_{\mathbf{yy}}^{-1} \mathbf{a}_k\|^2 \sigma_n^2} \qquad (6.5)$$

It can be shown that through spatial beamforming, up to M co-channel terminals can be perfectly separated in the absence of noise. The capability of simultaneous communications with multiple terminals using antenna array enables SDMA.

6.2.1 Conventional Spatial-Division Multiple-Access

Before we review the concept of SDMA, let us describe the packet radio network under consideration. The slotted packet network consists of K radio terminals actively communicating with a base station. The information is transmitted in blocks of symbols called *packets*. The time axis is divided into time *frames* of fixed length. Within each time frame, the spectral resource is evenly divided into L *slots* through which multiple packets can be transmitted. A slot is either a time slot, a frequency channel, or a spreading code among a set of orthogonal codes. Each terminal transmits packets during one of the L slots in a frame. The process repeats in following frames for all users.

The base station has total control of the network traffic. It is assumed that when a terminal wants to be admitted to the network, it places an admission-request packet through a reserved request slot, from which the base station obtains the spatial signature and data information of the terminal. The base station then assigns the terminal to certain data slots depending on the request and the traffic situation. The focus of this book is on how to schedule packet transmissions after the terminals are admitted and the part on access initiation is kept to its minimum.

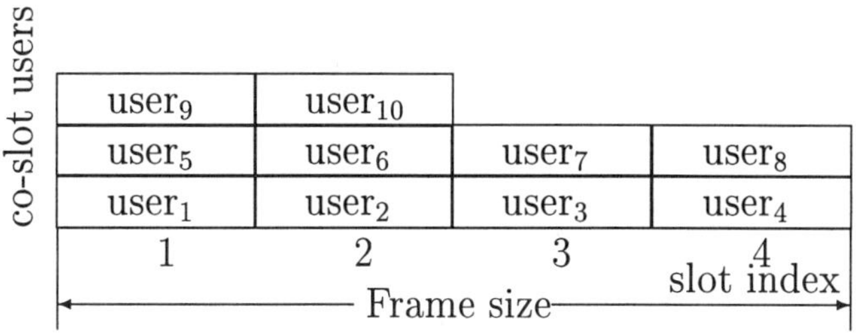

	1	2	3	4
co-slot users	user$_9$	user$_{10}$		
	user$_5$	user$_6$	user$_7$	user$_8$
	user$_1$	user$_2$	user$_3$	user$_4$

slot index ← Frame size →

Figure 6.2 SDMA without scheduling.

In most practical situations $K \gg M$. The SDMA accommodates these terminals in both the space and the time/frequency/code domain. Figure 6.2 illustrates how this objective is achieved in a 4-slot, 10-terminal SDMA network [6, 7]. Built upon the slotted packet network, the SDMA expands the capacity by allowing multiple (up to M) terminals in each slot, and within each slot spatial beamforming is performed to acquire packets from multiple terminals. In essence, SDMA adds another dimension to the spectrum resource by expanding each slot into M spatial slots (vertical axis), translating into an M-fold increase of system throughput in an ideal scenario.

In reality, however, the ability of capturing multiple packets depends critically on the spatial configuration of the co-slot terminals. Let $\{\mathbf{a}_k\}_{k=1}^{K}$ be the spatial signatures of the K terminals in the system. The terminals are distributed among the L slots in SDMA and:

$$\mathbf{A}_l = [\mathbf{a}_{l1} \cdots \mathbf{a}_{ld_l}]$$

denotes the array manifold of the co-slot terminals in the lth slot. The mapping between $\mathbf{a}_k$ and $\mathbf{a}_{ld_l}$ is one to one. The base station performs spatial beamforming on the received co-slot signals as described in Section 6.2. According to Poor and Verdu [10], the residual interference and noise of the MMSE beamforming output has Gaussian-like characteristics, allowing us to calculate the BER based on output SINR

in (6.5):
$$P_E = Q(SINR) \tag{6.6}$$

Assuming the error correction capability of a coded packet of N bits is t (i.e., a maximum number of t bits of error is correctable), the "packet success probability":

$$P_k = P(SINR_k) = \sum_{n=0}^{t} \binom{N}{n} Q(SINR_k)^l (1 - Q(SINR_k))^{N-n} \tag{6.7}$$

where $SINR_k$ is the beamforming output SINR of the kth terminal.

The *throughput* of each terminal in SDMA can be defined accordingly.

Definition *Let each terminal in SDMA transmit packets in one out of L slots, where L is the total number of slots in a fixed frame. The throughput of the k terminal is defined as:*

$$T_k = \frac{1}{L} P_k, \quad k = 1 \cdots, K \tag{6.8}$$

and the average throughput of the SDMA,

$$T_{ave} = \frac{1}{K} \sum_{k=1}^{K} T_k \tag{6.9}$$

Figure 6.3 illustrates the performance gain of SDMA by comparing the packet throughput in a 4-antenna, 36-terminal, 9-slot SDMA system with that of a single-antenna, 36-slot, TDMA system (only one terminal allowed in each time slot). The spatial signatures are generated as independent Gaussian vector (normalized to unit power). SDMA is fundamentally advantageous when compared to pure TDMA in that each terminal has four times the chance to transmit successfully. As a result, significant gains in throughput are obtained for all terminals. On the other hand, two shortfalls of conventional SDMA are evident:

1. The variation of individual terminals' performance is very large, primarily due to spatially aligned (or closed) terminals being in

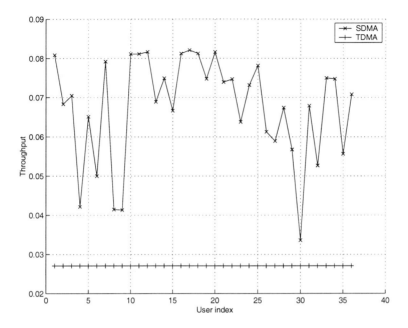

Figure 6.3 Throughput improvement of SDMA.

the same slot. Since the capability of most wireless systems is worst-case limited, the throughput improvement over pure TDMA is actually quite limited.

2. The example above assumes a proper number of terminals per slot. In real applications, the number of terminals having packets to transmit also varies significantly. If the number of slots in a frame is fixed, it can lead to too many or too few terminals in each slot. Either way, the gain will be significantly reduced.

Note that *optimum* beamforming is assumed in spatial separation, hence the above problems *cannot* be cured with physical layer treatment.

6.3 Scheduling Protocol

The objective of MAC-layer scheduling is to increase the efficiency of SDMA through intelligent MAC-level planning. In particular, the protocol (1) systematically assigns "most-orthogonal" terminals to the same time slot so that the worst case and average performance is optimized; (2) varies the total number of slots in a fixed frame so that that number of co-slot terminals is maintained at the level that is most suitable for spatial beamforming.

For simplicity, we invoke the following assumptions:

1. All terminals' spatial signatures are known to the base station.

2. All K terminals have packets to transmit. Each terminal transmits through only one out of the L slot.

3. All packets have the same priority.

The scheduling has two components: terminal allocation and slot adjustment. Before describing the details, we would like to point out that optimum scheduling by itself is an NP-complete combinatory problem whose solution can only be found with exhaustive searching, which is infeasible for most practical values of L and K. The ensuing design is hence most heuristic. Nevertheless, the protocol, as will be shown analytically and experimentally, demonstrates extremely high efficiency in the sense that it approaches the performance upper-bound in most scenarios.

6.3.1 Terminal Allocation

The objective of terminal allocations is to distribute the K terminals with spatial signature $\{\mathbf{a}_k\}_{k=1}^{K}$ into L slots so that the throughput of the resulting SDMA system is optimum. Such is accomplished as follows:

- At the beginning, the base station assigns the first L terminals to the L open slots. The terminal allocator keeps a record of the array manifold of each slot (after the first step, only one terminal in each slot):

$$\mathbf{A}_l = [\mathbf{a}_{l1} \cdots \mathbf{a}_{ld_l}], \quad l = 1 \cdots L$$

- For each additional terminal with spatial signature $\mathbf{a}_{new}$, the base station makes a channel-aware assignment based on a pre-determined criterion (described below). For example, using the network throughput as the design metric, the new terminal is assigned to the jth slot if it has the maximum average throughput after the new terminal is added:

$$T_{ave}([\mathbf{A}_j\ \mathbf{a}_{new}]) \geq T_{ave}([\mathbf{A}_l\ \mathbf{a}_{new}]), \quad l \neq j,\ l = 1 \cdots L$$

- After the new terminal is assigned to the kth slot, the base station updates the record of the assigned slot as follows:

$$\mathbf{A}_k = [\mathbf{A}_k\ \mathbf{a}_{new}]$$

- Steps 2 and 3 are repeated until all terminals are assigned.

Several criteria can be considered to make the assignment "channel-aware":

1. The mean-squared error (MSE) criterion: The new terminal is assigned to a slot that has the lowest average MSE after the new terminal is added. More specifically, the new terminal is assigned to slot l if:

$$\sum_{k=1}^{d_j+1} MSE_{j,k} \leq \sum_{k=1}^{d_l+1} MSE_{l,l} \quad j \neq l$$

where $MSE_{l,k}$ is the beamforming output MSE of terminal k in slot l with co-slot terminals $[\mathbf{a}_{l1} \cdots \mathbf{a}_{ld_l}\ \mathbf{a}_{new}]$

2. Throughput criterion: The new terminal is assigned to a slot that has a maximum average throughput after adding the new terminal:

$$T_{ave}([\mathbf{A}_k\ \mathbf{a}_{new}]) \geq T_{ave}([\mathbf{A}_j\ \mathbf{a}_{new}]) \quad j \neq k$$

The MSE criterion is easy to use and intuitively plausible: the new terminal is assigned to a slot with terminals (spatial signatures) most orthogonal to the new terminal (spatial signature). The criterion can be further simplified by using the average cross-correlation between co-slot terminals' spatial signatures. But as explained earlier, the beamforming performance is nonlinearly related to the covariance matrix of **A**. Its cross-correlation may not be a good indicator of final performance.

The throughput criterion requires more computations but meets the overall design objective in a packet network. The calculation of the throughput involves computing the SINRs based on the spatial signatures and translating the results into throughput using (6.5) to (6.9).

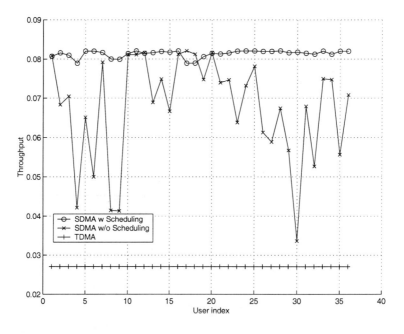

Figure 6.4 Throughput improvement in SDMA due to spatial signature-based allocation.

Figure 6.4 shows the effect of the above terminal allocation scheme with MSE as the allocation criterion. The system setup is the same

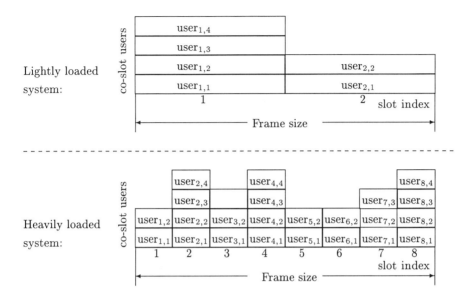

Figure 6.5 Channel-aware scheduling.

as that of Figure 6.3. With simple MAC-layer planning, the system throughput increases substantially across the board. More importantly, the variance of the terminals' throughput becomes so small that one can practically use the average throughput as the design parameter for the system capacity. This is particularly important to wireless applications whose capacity is usually worst-case limited.

6.3.2 Slot Adjustment

The efficacy of terminal allocation evidently depends on the number of slots in a frame. If the number of users is much higher (over-crowded slot) or lower (no co-slot terminals) than L, the throughput gain due to terminal allocation will disappear. For optimum scheduling, the protocol in principle adjusts the slot size so that as many co-slot terminals can be accommodated as possible without causing breakdown in throughput. In practice, this can be realized by merging multiple slots into one large slot. The idea of load-dependent slot adjustment is illustrated in Figure 6.5. The protocol expends the duration of the slot when the load of the system is light and reduces the slot size when the load is heavy. The actual way of assigning the terminals depends on the spatial configuration of all terminals. In practice, this is realized by merging multiple slots into a large-size slot.

To find the best L, the protocol simply evaluates all possible L values and finds the optimum frame partition. More specifically, the slot adjuster varies the L value and performs terminal allocation for all possible L values. The one that maximizes the predetermined metric (e.g., the average throughput) is chosen to the final number of slots per frame. Most often, L is limited to several values due to system constraints.

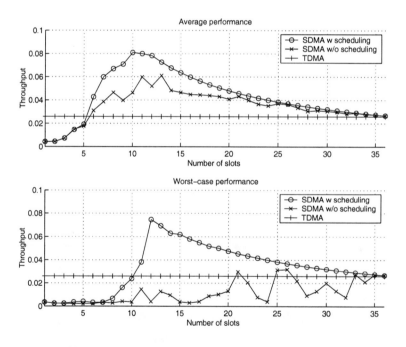

Figure 6.6 Throughput vs. number of slots.

Figure 6.6 shows the throughput for different L values given the setup in Figures 6.3 and 6.4. In this example, the curves indicate that the optimum average throughput is reached when $L = 10$, whereas the optimum worst-case throughput is reached when $L = 11$. In both cases, the average number of terminals per slot is slightly under $M = 4$, the dimension of the antenna array. Such verifies the importance of slot adjustment since in many cases (e.g., multiple terminals are spatial aligned), the optimum L value may be quite different from K/M.

In summary, scheduling improves the performance of SDMA by judiciously selecting two sets of variables based on the spatial configuration – the number of slots in each frame and terminals' slot indices. We shall refer to the protocol as "SDMA with scheduling" in the remainder of this book. In addition to the obvious gain in throughput, it is important to point out the SDMA with scheduling relies on *fixed* spatial operations and obtains the throughput without undue complexity during the access. Each terminal, after sending an access request, is given a (or multiple) slot assignment as in regular multiple-access mode. The scheduling process is carried by the base station without excessive information exchange between the terminals and base station.

The pseudocode description of the scheduling procedure is presented below. The inputs of the function are the spatial signature of the terminals and the outputs are the optimum partition of the frame (number of slots in a frame) and the terminal allocation.

```
Scheduler(a₁, a₂,···,a_K)
begin
      L=number of slots
      K=number of terminals
      a₁ ··· a_K= users' spatial signatures
      ARRAY_MANIFOLD[l]= lth slot's array manifold
      for each possible L value do
            for l = 1 : L, ARRAY_MANIFOLD[l]=a_l, end
            for k = L+1 : K
                  find l = 1 : L that T_ave(ARRAY_MANIFOLD[l]+a_k) is
maximum
                  ARRAY_MANIFOLD[l]=ARRAY_MANIFOLD[l] +a_k
            end
      end
      L_opt = L:  ∑_{l=1}^{L} T_ave(ARRAY_MANIFOLD[l]) is maximum
      return L_opt
      return ARRAY_MANIFOLD[l], l = 1···L_opt
```

176   Signal Processing Applications in CDMA Communications

end

The "SDMA with scheduling" scheme implicitly assumes that the traffic is steady (i.e., all terminals have packets to transmit at all times). In real applications, the number of terminals changes every time a terminal becomes active/inactive. Under these scenarios, it may not be reasonable to perform full-scale scheduling on all terminals every time there is a change in the network traffic. While protocol optimization for variable traffic patterns is an important topic that requires further investigation, one straightforward strategy is partial scheduling. That is, allocating only the newly active terminals based on their spatial signatures. The number of slots is maintained at $\lceil K/M \rceil$, where K is now the *average* traffic rate.

We shall term this protocol as "SDMA with partial scheduling."

6.4   Analysis

In this section, we analyze the efficiency of the SDMA with scheduling protocol by establishing its performance upper- and lower-bounds and investigating its behavior with an infinite population.

6.4.1   Upper-Bound and Lower-Bound

The following lemma defines the performance upper-bound of d co-slot terminals.

Lemma 1 *Let M be the number of antenna elements, d the number of terminals sharing one slot, $\{\mathbf{a}_i\}_{i=1}^{d}$, $\|\mathbf{a}_i\| = 1$ the spatial signatures associated with these terminals, and σ_n^2 the power of the additive white noise. The maximum MSE of the beamformer outputs is lower-bounded by:*

$$MSE_{max} \geq \begin{cases} \frac{\sigma_n^2}{1+\sigma_n^2} & d \leq M \\ \frac{d+(M-1)\sigma_n^2}{d+M\sigma_n^2} & d > M \end{cases} \qquad (6.10)$$

The equality holds when $\mathbf{A} = [\mathbf{a}_1 \cdots \mathbf{a}_d]$ are column orthogonal for the case of $d \leq M$, and row orthogonal for the case of $d > M$.

Proof:

Let MSE_{max} be the maximum MSE of the beamforming outputs,

$$MSE_{max} = MAX\,(MSE_1 \cdots MSE_d)$$

The case for $d \leq M$ is simple – MSE_{max} reaches its minimum when $\{\mathbf{a}_i\}_{i=1}^d$ are mutually orthogonal, that is, no multiterminal interference. It is straightforward to show that the optimum beamforming vector $\mathbf{w}_i = \mathbf{a}_i$ and the MSE value for each terminal is $\frac{\sigma_n^2}{1+\sigma_n^2}$.

When $d > M$, orthogonality between terminals' signature waveforms is no longer possible. Since:

$$MSE_i = 1 - \mathbf{a}_i^H \mathbf{R}_{\mathbf{yy}}^{-1} \mathbf{a}_i = 1 - \mathbf{a}_i^H \left(\mathbf{A}\mathbf{A}^H + \sigma_n^2 \mathbf{I}\right)^{-1} \mathbf{a}_i$$

to determine MSE_{max} we need to compute the minimum value of $(\mathbf{a}_i^H \mathbf{R}_{\mathbf{yy}}^{-1} \mathbf{a}_i)$ as a function of $\mathbf{A}$.

Let:

$$\lambda_1 \geq \lambda_2 \geq \cdots \geq \lambda_M$$

be the ordered eigenvalues of $\mathbf{R}_{\mathbf{yy}}$. Using properties of matrix operations, we have:

$$\frac{1}{\lambda_1} \leq \mathbf{a}_i^H \mathbf{R}_{\mathbf{yy}}^{-1} \mathbf{a}_i \leq \frac{1}{\lambda_M}$$

That is, $(\mathbf{a}_i^H \mathbf{R}_{\mathbf{yy}}^{-1} \mathbf{a}_i)_{min}$ is bounded by $\frac{1}{\lambda_1}$. The smaller λ_1 is, the larger $(\mathbf{a}_i^H \mathbf{R}_{\mathbf{yy}}^{-1} \mathbf{a}_i)_{min}$ is, and the smaller MSE_{max} is.

Since $\text{tr}(\mathbf{R}_{\mathbf{yy}}) = \sum_{i=1}^M \lambda_i$ is fixed, λ_1 reaches its minimum value when:

$$det(\mathbf{R}_{\mathbf{yy}}) = \prod_{i=1}^M \lambda_i$$

is maximum, which occurs when $\mathbf{A}\mathbf{A}^H = c\mathbf{I}$. That is, when the rows of $\mathbf{A}$ are mutually orthogonal [11].

To determine c, we notice that:

$$c = \frac{1}{M}\text{tr}(\mathbf{A}\mathbf{A}^H) = \frac{1}{M}\text{tr}(\mathbf{A}^H\mathbf{A}) = \frac{d}{M}$$

Thus, $\mathbf{R_{yy}} = (\frac{d}{M} + \sigma_n^2)\mathbf{I}$ and:

$$MSE_{max} = \frac{d + (M-1)\sigma_n^2}{d + M\sigma_n^2}$$

Since the packet throughput is proportional to the output SINR and inverse proportional to the MSE, it is not difficult to calculate the throughput bound for co-slot terminals. For most practical cases where the signal-to-noise ratio is high, co-slot interference is the limiting factor of throughput and the worst-case throughput drops significantly when the slot is overloaded (i.e., more terminals than the number of antennas in each slot). Hence, in the discussion below, we shall limit our attention to the cases when $d \leq M$.

With Lemma 1, we now consider the case of SDMA with variable slot length.

Theorem 1 *For a K-terminal, M-antenna system, the worst-case throughput upper-bound and lower-bound are given by:*

$$\frac{1}{K}P(1/\sigma_n^2) \leq T_{wrst} \leq \frac{1}{L}P(1/\sigma_n^2), \quad L = \lceil \frac{K}{M} \rceil$$

Proof:

The lower-bound can be easily reached by partitioning the frame into K slot (pure TDMA), with each terminal only using one of the slots.

The upper-bound corresponds to the case when K terminals are distributed into $L = \lceil \frac{K}{M} \rceil$, with no slot having more than M terminals, and all terminals in the same slot are mutually orthogonal.

6.4.2 Asymptotic Optimality

When the number of users to be scheduled becomes larger, it is more likely for the scheduler to group orthogonal terminals together and yield better performance. The performance of the scheduler for a 2-antenna system is given by the following theorem:

Theorem 2 *For a two-antenna system, the average throughput of SDMA with scheduling reaches the throughput upper-bound for an infinite terminal population.*

Proof:

Assume there are K terminals, with 2×1 spatial signatures $\{\mathbf{v}_k\}$ uniformly distributed on half-unit circle $[0, \pi]$. The scheduler assigns these terminals into slots, with each slot having a maximum of two terminals.

To show the optimality, let us equally divide $[0, \pi/2]$ into $L = \lceil \frac{2\pi}{\epsilon} \rceil$ regions, denote terminals as: $P_1^1, P_1^2, \ldots, P_1^L$. Similarly divide $[\pi/2, \pi]$ into $P_2^1, P_2^2, \ldots, P_2^L$. Assume there are i terminals with spatial signatures in P_1^k and j terminals with spatial signatures in P_2^k. Then the distribution of i and j can be calculated as:

$$p(i,j) = \binom{K}{i+j}\binom{i+j}{j}(1-\frac{1}{L})^{K-i-j}(\frac{1}{L})^{i+j}\frac{1}{2}^{i+j} \qquad (6.11)$$

If $\mathbf{v}_m \in P_1^k$ and $\mathbf{v}_n \in P_2^k$ (i.e., they are almost orthogonal), then the scheduler will assign them into one slot. The cross-correlation between these two terminals is:

$$\rho_{paired} = \mathbf{v}_n^T \mathbf{v}_m \leq sin(\pi/L) \leq \epsilon/2 \qquad (6.12)$$

If one terminal in P_1^k cannot find a pair in P_2^k or vice versa, then the cross-correlation $\rho_{unpaired} \leq 1$. The average cross-correlation over all terminals can be calculated as:

$$\begin{aligned}
\mathbf{E}\{\rho\} &= \mathbf{E}\{\tfrac{2min(i,j)\rho_{paired}}{i+j} + \tfrac{|i-j|\rho_{unpaired}}{i+j}\} \\
&\leq \mathbf{E}\{\tfrac{min(i,j)\epsilon}{i+j} + \tfrac{|i-j|}{i+j}\} \\
&\leq \epsilon/2 + \mathbf{E}\{\tfrac{|i-j|}{i+j}\} \\
&\leq \epsilon/2 + \mathbf{E}\{\mathbf{E}\{|i-j| | i+j=n\}/n\} \\
&= \epsilon/2 + \mathbf{E}\{\mathbf{E}\{|i-j| | i+j=n\}/n\} \\
&\leq \epsilon/2 + \mathbf{E}\{\sqrt{\mathbf{E}\{|i-j|^2 | i+j=n\}}/n\} \\
&= \epsilon/2 + \mathbf{E}\{\sqrt{\mathbf{E}\{4i^2 - 4ni + n^2\}}/n\} \\
&= \epsilon/2 + \mathbf{E}\{\sqrt{n}/n\} \\
&= \epsilon/2 + \mathbf{E}\{\sqrt{1/n}\} \\
&\leq \epsilon/2 + \mathbf{E}\{\sqrt{2/(n+1)}\} \\
&= \epsilon/2 + \sqrt{2}\sqrt{\mathbf{E}\{1/(n+1)\}} \\
&\leq \epsilon/2 + \sqrt{L/K}
\end{aligned} \qquad (6.13)$$

Clearly, when $K \geq \frac{8\pi}{\epsilon^3}$, $\mathbf{E}(\rho) \leq \epsilon$. That is, when $K \to \infty$, $\rho \to 0$. In other words, when K goes to infinity, by scheduling, terminals in most slots are almost orthogonal, and the average throughput approaches the upper-bound.

A rigorous proof for the $M > 2$ case is quite involved because of the difficulty associated with partitioning a sphere in a high-dimensional space. However, extensive simulations show the throughput upper-bound is reached when the number of terminals is large regardless of M.

Figure 6.7 demonstrates the asymptotic performance of SDMA with scheduling. In this particular setup with 4 antennas, the total throughput upper-bound per time-spatial slot is 0.92. Without scheduling, the throughput is at about 0.64. By applying scheduling, the throughput increases to 0.75 with just 16 terminals. For 32 terminals, the throughput is more than 0.82, and when the number of terminals is more than 100, the throughput is quite close to the upper-bound. This shows that the scheduling works well even with a small number of terminals, although the performance is optimum when the number of terminals is large.

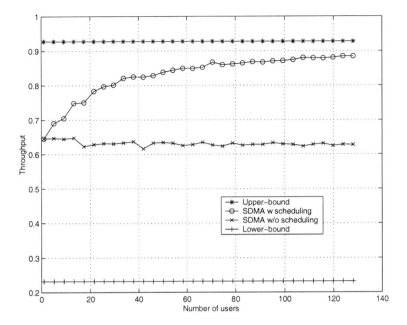

Figure 6.7 Asymptotic performance of SDMA with scheduling.

6.5 Performance Evaluation

Our analysis so far has been based on the premise that all terminals have packets to transmit at all times, which is not always true in practical wireless networks [12]. In this section, we present some performance results for a network exploiting spatial diversity with more realistic traffic models. The SDMA schemes (without scheduling, with scheduling, and with partial scheduling) are studied under two types of data traffic models.

6.5.1 Finite Population With Bursty Traffic

For a finite population and bursty traffic [13], we assume:

1. Fixed number of users in the system, each having a unique spatial signature.

2. The packets arrive at each terminal in Poisson distribution with

rate λ when this terminal is in the ON period. No packet arrives in the OFF period. The length of the ON and OFF periods obeys certain distributions.

3. The spatial signature of each terminal varies slowly in comparison with the length of the ON period, so that all the packets arriving during one period have the same spatial signature.

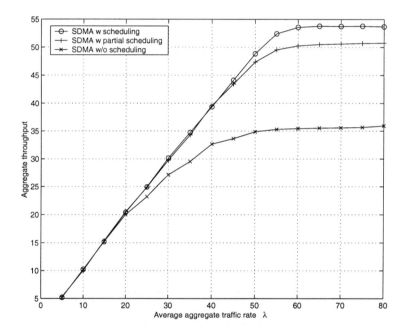

Figure 6.8 Throughput, burst traffic.

The above case presents a realistic WLAN type of scenario that is most suitable for SDMA [2]. In practice, once a terminal becomes ON, it submits a request to the scheduler, who marks the terminal active and estimates its spatial signature. It then either randomly assigns one slot to this terminal (SDMA without scheduling) or allocates a slot based on its spatial signature. The terminal, after receiving the scheduling information, begins to transmit the backlogged packets at the given slot.

We simulate a base station with 4 antennas serving 256 terminals. The length of the OFF period is exponentially distributed with mean time 1024. The ON period has fixed length 8. During the ON period, packets arrive according to the Poisson process at rate 2.5. The average aggregate traffic rate λ is given by:

$$\lambda = \frac{\text{Traffic Rate during ON period} \times \text{Number of Users} \times \text{ON Period}}{\text{ON Period} + \text{OFF Period}}$$

For SDMA without scheduling, the number of slots in a frame is fixed at 16. The noise power is set at a level such that the packet success rate for pure TDMA is 0.95.

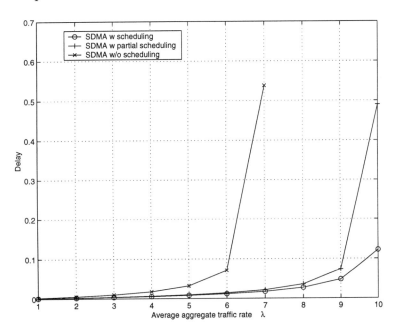

Figure 6.9 Delay, bursty traffic.

Figure 6.8 and Figure 6.9, respectively, show the throughput and delay of the three SDMA schemes as functions of the aggregate traffic rate λ. For low traffic, the rise in throughput is approximately linear with the traffic rate for both schemes. In this region, there is essentially no collisions due to failure in beamforming acquisition. As the traffic increases, the effect of scheduling becomes more and more evident. The throughput of SDMA with scheduling saturates after the traffic rate reaches 60, showing a throughput gain of almost 60 percent over that of conventional SDMA (without scheduling). The improvement in delay characteristics due to scheduling are as significant as that in throughput. As noticed from Figures 6.8 and 6.9 there is only a little performance gap between the full scheduling scheme and the partial scheduling scheme. This is particularly encouraging for practical applications that require minimum system overhead.

6.5.2 Infinite Population With Poisson Traffic

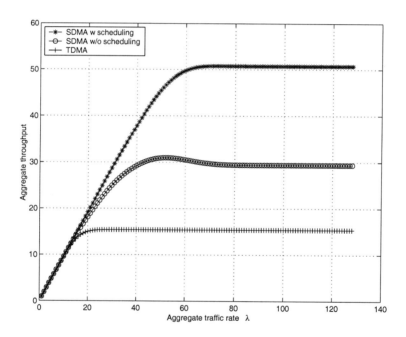

Figure 6.10 Throughput, data traffic.

The infinite population model is an analogy to the typical idealized slotted multiaccess model [14]. We have the following assumptions:

1. Packets arrive at each terminal according to independent Poisson processes, with λ being the overall arrival rate to the system, and λ/m the arrival rate at each transmitting node.

2. Each terminal has its own spatial signature.

3. The system has an infinite set of terminals and each newly arriving packet is from a new terminal.

4. The scheduler has the exact knowledge of the status of the terminal and its spatial signature.

Assumption 3 indicates a rapidly changing wireless network where the spatial signature of each terminal varies between two consecutive packets. Assumption 4 is not always possible, especially in rapidly changing wireless networks. However, the results derived under this assumption will provide insight into the average performance of the SDMA system.

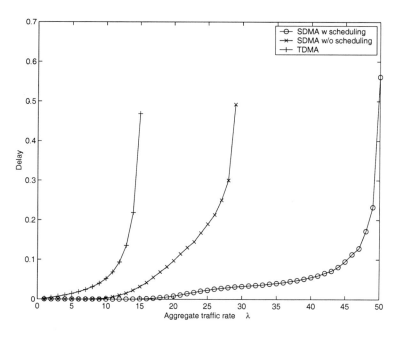

Figure 6.11 Delay, data traffic.

To evaluate the throughput and delay performance, we again simulate a network with 4 antennas and a frame with fixed 16 time slots for SDMA without scheming. The single-user packet success rate is 0.95.

Figure 6.10 and Figure 6.11, respectively, compare the throughput and delay characteristics for regular TDMA, SDMA without scheduling, and S-TDMA with scheduling as functions of the aggregate traffic rate λ. We can see that by exploiting the spatial diversity, the performance gain over pure TDMA is significant. An additional 60 percent improvement in throughput is obtained by MAC-layer scheduling. The curves for infinite population exhibit similar behavior to that of finite population with bursty traffic. Since these two cases represent the two extremes of the mobile wireless network – high-speed/rapidly changing environment and low-speed/near-stationary setup – it is reasonable to conclude that SDMA with scheduling can perform with high efficiency in a variety of scenarios.

6.6 Conclusion

In this chapter, we described a scheduling scheme for space-division multiple-access to improve the performance of a slotted antenna array packet network. The protocol judiciously selects co-slot terminals based on their spatial signatures to enable significant improvement in the system throughput-delay characteristics. We analyzed the performance of the protocol by deriving its throughput upper- and lower-bounds and establishing its asymptotic optimality. Simulation results showed that the new scheme is robust with respect to traffic pattern, even with partial adaptation where only the new terminal is scheduled.

An ideal application of the SDMA with scheduling protocol is the wireless local-area network. In a WLAN environment, (1) the spatial configuration is relatively stable; (2) unlike other wireless networks such as the PCS, WLANs are local, use unlicensed bands, and are subject to much fewer regulations; (3) the latest IEEE 802.11a modem is OFDM-based, which reduces wideband spatial processing into parallel narrowband beamforming. For these reasons, the protocol described has the potential to make an immediate impact on next-generation WLANs.

References

[1] D. J. Goodman, S. Nanda, and U. Timor, "Performance of PRMA: A packet voice protocol for cellular systems," *IEEE Trans. on Vehicular Technology*, **40**, August 1991.

[2] Zhao Liu, Mark J. Karol, and Kai Y. Eng, "Distributed-queueing request update multiple access (DQRUMA) for wireless packet (ATM) networks," In *IEEE Proc. ICC*, pages 1224–1231, Seattle, WA, 1992.

[3] Alex E. Brand and A. Hamid Aghvami, "Multidimensional PRMA with prioritized Bayesian broadcast – A MAC strategy for multiservice traffic over UMTS," *IEEE Trans. on Vechicular Technology*, **47**, November 1998.

[4] Jim Geier, *Wireless LANs*, Macmillan Technical Publishing, U.S.A., 1999.

[5] S. Kumar and D. R. Vaman, "An access protocol for supporting multiple classes of service in a local wireless environment," *IEEE Trans. on Vehicular Technology*, **45**(2), May 1996.

[6] G. Xu and S. Q. Li, "Throughput multiplication of wireless LANs for multimedia services: SDMA protocol design," In *Proc. Globecom*, San Francisco, CA, November 1994.

[7] J. Ward and R. T. Compton, "High throughput slotted ALOHA packet radio networks with adaptive arrays," *IEEE Trans. Communications*, **41**(3):460–470, March 1993.

[8] Jean-Paul M. Linnartz, "Imperfect sector antenna diversity in slotted ALOHA mobile network," *IEEE Trans. on Communications*, **44**(10):1322–1328, October 1996.

[9] D. H. Johnson and D. E. Dudgeon, *Array Signal Processing: Concepts and Techniques*, Englewood Cliffs, NJ: Prentice Hall, 1993.

[10] H. V. Poor and S. Verdu, "Probability of error in MMSE multiuser detection," *IEEE Trans. on Information Theory*, **IT-43**(3):858–871, May 1997.

[11] B. Suard, G. Xu, H. Liu, and T. Kailath, "Channel capacity of space-division multiple-access schemes," *IEEE Trans. on Information Theory*, 1999.

[12] Yuguang Fang and Imrich Chlamtac, "Teletraffic analysis and mobility modeling of PCS networks," *IEEE Trans. on Communications*, **47**, July 1999.

[13] D. A. Levine, I. F. Akyildiz, and Inwhee Joe, "A slotted CDMA protocol with BER scheduling for wireless multimedia networks," *IEEE/ACM Trans. on Networking*, **2**, April 1999.

[14] Dimitri Bertsekas and Robert Gallager, *Data networks*, Prentice Hall, Inc., Upper Saddle River, NJ, 1992.

About the Author

Hui Liu received a B.S. in 1988 from Fudan University, Shanghai, China; an M.S. in 1992 from Portland State University, Portland, Oregon; and a Ph.D. degree in 1995 from the University of Texas at Austin, all in electrical engineering. From June to December 1996, he served as the director of engineering at Cwill Telecommunications, Inc., where he played a key role in the design and commercialization of the first smart antenna S-CDMA wireless local loop system. Dr. Liu held the position of assistant professor at the Department of Electrical Engineering at University of Virginia from September 1995 to July 1998. He is currently with the Department of Electrical Engineering at the University of Washington, Seattle. His research interests include broadband wireless networks, array signal processing, DSP and VLSI applications, and multimedia signal processing. He has published more than 25 journal articles and has six awarded or pending patents.

Dr. Liu's activities for the IEEE Communications Society include membership on several technical committees and serving as an editor for the IEEE Transactions in Communications. He is a recipient of the 1997 NSF CAREER Award.

Index

aggregate traffic rate, 183
antenna array, 19, 55, 164
    array manifold, 165
    packet network, 164

bandwidth expansion, 4
blind estimation, 50
    2D RAKE receiver, 90
    MC-CDMA, 135
    overloaded system, 55
bursty traffic, 181

carrier offset, 61, 128
    analysis, 141
    Guassian distribution, 145
    uniform distribution, 145
CDMA, 7
    aperiodic, 13
    generic, 10
    multicarrier, 121
    narrowband, 14
    periodic, 13
    receiver, 11
    synchronous, 14
    wideband, 16
cellular telephony, 1
channel-aware scheduling, 170
closed-form estimation, 53
coherent combiner, 86

constrained MOE, 27
    adaptive, 37
    asymptotic performance, 103
    downlink, 103
    equalization, 103
    steady state, 41
    step size, 110
cyclic prefix, 123

decorrelating RAKE receiver, 32
    adaptive implementation, 37
    steady state behavior, 41
delay spread, 14
deterministic least-squares, 96
direction-of-arrival, 27
Doppler shift, 50
downlink, 9, 101
    data model, 101
    equalization, 103
downlink equalizer, 101
    adaptive implementation, 107
    constrained MOE, 103
    convergence analysis, 109
    step size, 110

effective number of channels, 84
effective signature waveform, 30
estimation, 50
    blind, 90

192 *Signal Processing Applications in CDMA Communications*

carrier offset, 61, 136
channel, 51, 135
joint, 61
MC-CDMA, 135

flat-fading channel, 14
frequency-hopping, 6
frequency-selective channel, 14, 16

guard period, 29, 123

interchip interference, 17
intersymbol interference, 17, 29
ISI-free CDMA, 29

long codes, 13

matched filter, 5
MC-CDMA, 121, 126
   carrier offset estimation, 135
   carrier sensitivity, 141
   channel estimation, 135
   modulator, 127
   receiver, 129
      coherent combiner, 129
      MMSE receiver, 129
   spectrum, 128
MC-DS-CDMA, 130
   enabling codes, 137
   modulator, 130
   receiver, 132, 137
MCM, 122
medium-access control, 162
   scheduling, 170
multi-input multi-output, 84
multiple-access, 7

CDMA, 9
FDMA, 8
SDMA, 8, 161, 166
TDMA, 8
Multitone CDMA, 134
   modulator, 134
   spectrum, 134
multiuser detector, 11
MUSIC, 53, 135, 136

narrowband, 14
null spectrum, 67

OFDM, 123
   channel effect, 125
   cyclic prefix, 124
   modulator, 122
   power spectrum, 124
optimum 2D RAKE receiver, 87
orthogonal subspace, 52
oversampling, 84, 101
   spatial, 84
   temporal, 84

packet networks, 164, 166
   frame, 166
   scheduling, 166
   slot, 166
PCS, 1
performance analysis, 38, 56, 69, 106, 176
   first-order perturbation, 56, 69
Poisson traffic, 184
polynomial matrix, 65
   null, 66
   projection, 66

Index

principal component, 91
pseudorandom sequence, 4

RAKE receiver, 12, 17
    2D, 85
    decorrelating, 37
    performance comparison, 90
receiver
    coherent combiner, 86, 129
    downlink, 101
    MMSE, 11, 87, 129
    multiuser, 11
    single-user, 12
root distribution, 67

scheduling, 170
    allocation, 170
    asymptotic optimality, 178
    channel-aware, 170
    lower-bound, 178
    performance, 181
    slot adjustment, 173
    upper-bound, 178
SDMA, 8, 161
    scheduling, 170
    throughput, 167
short codes, 13
signal subspace, 52
signature waveform, 5
    effective, 16, 30
single-input multi-output, 101
single-user detector, 12
Smith form, 76
spatial diversity, 19
spatial signature, 20, 165
spread spectrum, 3

direct-sequence, 3
frequency-hopping, 6
modulation, 4
subspace decomposition, 52
SVD, 52, 136

throughput, 167
    SDMA, 167
throuhgput
    SDMA
        lower-bound, 176
        upper-bound, 176
traffic, 181
    bursty, 181
    Poisson, 184

unimodular polynomial matrix, 76
uplink, 9, 83

Walsh, 16, 139
    matrix, 16
wireless LAN, 1, 186

Recent Titles in the Artech House Mobile Communications Series

John Walker, Series Editor

Advances in Mobile Information Systems, John Walker, editor

An Introduction to GSM, Siegmund M. Redl, Matthias K. Weber, and Malcolm W. Oliphant

CDMA for Wireless Personal Communications, Ramjee Prasad

CDMA RF System Engineering, Samuel C. Yang

CDMA Systems Engineering Handbook, Jhong S. Lee and Leonard E. Miller

Cell Planning for Wireless Communications, Manuel F. Cátedra and Jesús Pérez-Arriaga

Cellular Communications: Worldwide Market Development, Garry A. Garrard

Cellular Mobile Systems Engineering, Saleh Faruque

The Complete Wireless Communications Professional: A Guide for Engineers and Managers, William Webb

GSM and Personal Communications Handbook, Siegmund M. Redl, Matthias K. Weber, and Malcolm W. Oliphant

GSM Networks: Protocols, Terminology, and Implementation, Gunnar Heine

GSM System Engineering, Asha Mehrotra

Handbook of Land-Mobile Radio System Coverage, Garry C. Hess

Handbook of Mobile Radio Networks, Sami Tabbane

High-Speed Wireless ATM and LANs, Benny Bing

Introduction to Mobile Communications Engineering, José M. Hernando and F. Pérez-Fontán

Introduction to Radio Propagation for Fixed and Mobile Communications, John Doble

Introduction to Wireless Local Loop, William Webb

IS-136 TDMA Technology, Economics, and Services, Lawrence Harte, Adrian Smith, and Charles A. Jacobs

Mobile Communications in the U.S. and Europe: Regulation, Technology, and Markets, Michael Paetsch

Mobile Data Communications Systems, Peter Wong and David Britland

Mobile Telecommunications: Standards, Regulation, and Applications, Rudi Bekkers and Jan Smits

Personal Wireless Communication With DECT and PWT, John Phillips and Gerard Mac Namee

Practical Wireless Data Modem Design, Jonathon Y. C. Cheah

Radio Propagation in Cellular Networks, Nathan Blaunstein

RDS: The Radio Data System, Dietmar Kopitz and Bev Marks

Resource Allocation in Hierarchical Cellular Systems, Lauro Ortigoza-Guerrero and A. Hamid Aghvami

RF and Microwave Circuit Design for Wireless Communications, Lawrence E. Larson, editor

Signal Processing Applications in CDMA Communications, Hui Liu

Spread Spectrum CDMA Systems for Wireless Communications, Savo G. Glisic and Branka Vucetic

Understanding Cellular Radio, William Webb

Understanding Digital PCS: The TDMA Standard, Cameron Kelly Coursey

Understanding GPS: Principles and Applications, Elliott D. Kaplan, editor

Universal Wireless Personal Communications, Ramjee Prasad

Wideband CDMA for Third Generation Mobile Communications, Tero Ojanperä and Ramjee Prasad, editors

Wireless Communications in Developing Countries: Cellular and Satellite Systems, Rachael E. Schwartz

Wireless Technician's Handbook, Andrew Miceli

For further information on these and other Artech House titles, including previously considered out-of-print books now available through our In-Print-Forever® (IPF®) program, contact:

Artech House
685 Canton Street
Norwood, MA 02062
Phone: 781-769-9750
Fax: 781-769-6334
e-mail: artech@artechhouse.com

Artech House
46 Gillingham Street
London SW1V 1AH UK
Phone: +44 (0)20 7596-8750
Fax: +44 (0)20 7630-0166
e-mail: artech-uk@artechhouse.com

Find us on the World Wide Web at:
www.artechhouse.com